AVATARS OF THE WORD

AVATARS OF THE WORD

From Papyrus to Cyberspace

JAMES J. O'DONNELL

Harvard University Press
Cambridge, Massachusetts
London, England

Printed in the United States of America
Fourth printing, 2000

First Harvard University Press paperback edition, 2000

Library of Congress Cataloging-in-Publication Data

O'Donnell, James Joseph, 1950–
Avatars of the word : from papyrus to cyberspace /
James J. O'Donnell.
p. cm.
Includes bibliographical references and index.
ISBN 0-674-05545-4 (cloth)
ISBN 0-674-00194-X (pbk.)
1. Communication and technology—History.
2. Written communication—History.
3. Cyberspace.
I. Title.
P96.T42036 1998
302.23′09—dc21 97-32256

To Ann

Aunt Jennifer's tigers prance across a screen,
Bright topaz denizens of a world of green.
They do not fear the men beneath the tree;
They pace in sleek chivalric certainty.

—Adrienne Rich

CONTENTS

PREFACE

This book is for people who read books and use computers and wonder what the two have to do with each other. We may sometimes think we are back at that moment in Victor Hugo's *Hunchback of Notre Dame* when a character brandishes a new-fangled printed book in his hand against the backdrop of the cathedral and exclaims "This will kill that!" For us, the question is whether the computer will put the book out of business. And if so, what becomes of us readers?

Of course, if you turn from this book to a person sitting near you and ask this question you will in a way be answering it. The spoken word has been not supplanted by the written or the printed word, but supplemented. Each new generation of technological advance adds to the possibilities and makes the interplay among different media more complex. We lose and we gain at the same time.

This book is an attempt to think about how we rework some of the connections among speaking and writing and reading today. It offers a historical perspective based in western cultures from Greco-Roman antiquity to the present, but it is not a history. Histories of the written word may be found elsewhere and I have not tried to duplicate them. Rather, this book is an exploration of what those histories have to say about us, the pasts we receive, and the futures we shape.

The first five chapters offer meditations on historical situations designed to suggest ways of thinking about our own times, and the last four explore the state of learning and teaching today. These chapters run in roughly chronological order from ancient to contemporary times; but they take up issues that have significance over wide spans of time, and so each discussion of ancient, medieval, and early modern events leads to different contemporary implications. Perhaps most unusual, there is regular recourse throughout the book to a point of view based in an unfamiliar neighborhood: Latin late antiquity.

I assume here that Latin late antiquity, the world from roughly 300 C.E. to 600 C.E., offers a distinct and useful vantage point from which to consider the development of our ways of recording, using, and transmitting the written word. To an extent blurred by conventional narratives of ancient, medieval, and modern times, late antiquity was a point of departure for a wide range of institutions still in current use. Churches, law courts, schoolrooms, and libraries reflect the innovations of that age, while Europe's boundaries still eerily reflect the map of the later Roman empire. My references to the shaping writers and practices of that time offer a vantage point that, to be sure, reflects my own professional competence as a scholar, and I will not claim that it is the best choice simply because it is mine. Let the reader judge how useful it is.

The style of these chapters is deliberately associative and informal. While no single line of argument is advanced, the book is structured as an increasingly focused series of meditations approaching the issues and experiences of our own time. Some shorter discussions, which I call "hyperlinks," appear between chapters, interrupting the main line of discussion to expand on some points of interest.

That this book is exceedingly personal, even familiar, is by design. Whatever abstract ideas we have about what the past means to us and how it works in our culture, we will understand its real value best if we attend to how that past works in individual present lives. What is it that our past does for us? Or rather, what is it that the common construction of the past does for us?

The question is as apposite of remote antiquity as it is of the small child on a public conveyance I heard not long ago, repeatedly trying to break into her parents' conversation until she got a chance to remind them of the last time they had been to this touristy place—and then got, at precocious age, a wistful faraway sound in her voice when she remembered a detail of that event that had so far escaped her. Why did she need to remember that and talk about it? Where did she learn that wistfulness? Public history and private history can be powerful beyond all reasonable measure.

For now, I tell my own story to make a first effort at that kind of specificity. The reader of this book will no doubt have similar autobiographical reflections, not always as sober and respectable as we would like them to be. Our culture's heritage is the cumulative total of what that heritage means to all the individuals in it, not a cerebral abstraction of what it ought to be like or could be like. But, as I will show, that vision of ought and could is itself an influential factor in shaping people's expectations of themselves, and shaping their allegiances and their rebellions as well.

Finally, though this is a traditional book in hard covers, it is not without its own electronic avatar,* a homepage all its own, to which I refer the reader for materials, especially illustrations, that supplement the printed book in numerous ways. There will be World Wide Web addresses scattered throughout the book to point to specific items, but

*I take "avatar" throughout in the sense of "manifestation"—the form in which some abstract and powerful force takes palpable shape for human perception.

if the curious reader should venture to *http://ccat.sas.upenn.edu/jod/avatars,* an organized approach to all the materials collected there for this book is available.

Those who have heard me speak or rummaged in my web site will have seen this book looming into view for some years now. The dedication expresses my deepest indebtedness, but apart from that I must limit myself to a generic expression of gratitude to all those who have invited me to join their conversations during these exciting years, even if I have done so only briefly and from the fortified position of a lecture hall podium.

AVATARS OF THE WORD

Jerome in His Study (with Lion)

Introduction

THE SCHOLAR IN HIS STUDY

The fifteenth-century portrait of St. Jerome facing this page famously embodies a familiar image: the saint with his lion in a spacious study, poring over his books.* To a certain taste, the appeal of this scene is an irresistible model of the scholarly life: well-chosen books, seemly surroundings, dignity, and tranquility. When a classically trained scholar looks at this image, it can be hard to suppress the temptation to see oneself in Jerome's robes. Our domestic architecture, habitual apparel, and choice of companions may differ from his, but our powers of self-aggrandizement let us indulge for a moment the fantasy. The historical Jerome, irascible and charming by turns, settled in Bethlehem in the last years of the fourth century C.E. and devoted himself to translating scripture from Hebrew and Greek into Latin, while carrying on a self-advertising correspon-

*The theme continues, perhaps unconsciously: "A work in progress quickly becomes feral . . . it is a lion you cage in your study. As the work grows it gets harder to control . . . You must visit it every day and reassert your mastery over it. If you skip a day, you are, quite rightly, afraid to open the door to its room. You enter with bravura, holding a chair at the thing and shouting, 'Simba!' " (Annie Dillard, *The Writing Life* [New York, 1989] 52).

dence with social and literary eminences across the Roman world. By the time of the Renaissance he had become an object of veneration and a model to imitate, and had been reshaped in the process; the image and some traces of the veneration still resonate in us.

When we indulge the fantasy of identification with Jerome, we engage in an old exercise. There is evidence to suggest that the moment of first unveiling of this very painting was meant to encourage exactly such a collapsing of present and past. Though the image is clearly intended to be "Jerome," the details of his clothing have attracted attention. At the time of this painting, only two men were entitled to wear this costume of papal legate and thus were possible models for the portrait. One was an elderly man of no standing whatever, and the other was the polymath diplomat Nicholas of Cusa, whose works inspire to this day a learned society and a cottage industry of study.

Such an intermingling of present and past was very much in the air in the Renaissance and remains a familiar habit of mind today. Erasmus, for example, created and shaped his public image in the early decades of the sixteenth century by modeling it expressly on that of the church fathers of a thousand years before. He used the technology of print to make his own scholarly practice seem old and authoritative, though he was a man modern down to his toes and an adventurous entrepreneur of the lately invented printed book. One of Erasmus' achievements as a scholar was a biography of Jerome based for the first time not on hagiographic tradition but on a chronologically assessed and digested reading of Jerome's letters. When we look at the immense modern edition of Erasmus' letters in that light, we are unconsciously letting him teach us how to read them just as he would have us read Jerome, as transparent historical evidence that will lead us to the judgment of him that he wanted us to make. Ironically, another recent study shows how Jerome himself created the literary and public role he took in the late antique world by conscious imitation of the third-century Christian philologist Origen. Erasmus and Jerome were their own first image managers.

Such exercises in authorizing the present out of the past can be quite effective. Erasmus was a mildly renegade monk of no social standing, a pious Christian of deep if mildly idiosyncratic views, and a man who spent a great deal of time and effort in evading the polemic that shattered the religious unity of Europe in his time (while getting in a few shots of his own). But he succeeded beyond even surely his own dreams in creating an image of the warm, friendly, accessible, humane, and reasonable scholar that has found favor far beyond the circles his professed ideas about God and man could reasonably have been expected to penetrate. His friend and contemporary Thomas More had a long run as a nearly equally successful purveyor of an image of himself as benevolent humanist, but his most recent biographer has finally succumbed to the twentieth century's drive to debunk. The glowing portrait from *A Man for All Seasons* is now shadowed by the image of heretic-hunter. But Erasmus survives because he succeeded in substituting the image for the man. The rebarbativeness of some of his opinions disappears behind that image.

To rehearse the principled objections that can be raised against this kind of costume party self-presentation will do no more than highlight some of the effects of a practice widely employed in every age. Success breeds imitation, and ambition accordingly refines the art. I will single out only a few implications of this way that scholars in particular have of seeing themselves in a tradition.

Sitting at a keyboard before a computer screen, surrounded by mass-produced printed books, proud holder of a library card that provides me access to millions of books any day I like, and author of several published books of my own that are stored in that same repository, I am in many respects a very different person from an Erasmus or a Nicholas of Cusa. Erasmus knew the printed book and scrambled to take advantage of it, and Nicholas's secretary was one of the first entrepreneurs of printing in Italy decades before Erasmus, but neither of them could properly imagine a world so awash in books as ours is. For them any book was still a precious thing, and learning of any kind a struggle against the outright disappearance or inaccessibility of the words of the past and present day. (In Erasmus' time, people would still make handwritten copies of printed books, just to have a copy.)

A Renaissance painter constructing an image of Jerome was equally anachronistic, for Jerome lived again in very different conditions, in a world where the codex book was a relative novelty, and where the Christianity he fought for was just barely learning how to manage a library that contained more than scripture. Even scripture was only patchily available in translations of indifferent quality, and Jerome's own greatest achievement would be to put into circulation a better and more consistent Latin version of the Bible for his time. By the fifteenth century, the question was how to try to distribute universally the thing that Jerome had labored to make possible at all.

These inscriptions of ourselves into the past, these revivals of the past in ourselves, are distinctly ahistorical in many ways, but particularly in the way they blur together conditions of learning and language that are radically different. Jerome once ran across a Greek word in a text, and wrote to a friend that he remembered seeing that word only twice elsewhere, once in scripture, once in an apocryphal religious work. As it happens, he was correct: the three passages he knew are the only places (still) where we know that word to have been used in the written legacy of Greek literature. Hearing that story, I marvel at the powers of Jerome's memory, knowing that as a modern scholar with some similar interests in scripture and translation, I would never dare to say such a thing. I attribute this to the distractedness of my education, as well as my inability to read and retain everything that I would like to, but, at bottom, I have a suspicion that in those days people trained their memories to be better than ours are and that weakling reliance on the printed word has sapped our powers of memory.

Another way of looking at it is to say that Jerome's advantage over me lies in the emptiness of his textual memory, not its fullness. He did not have whole ranges of synapses cluttered with lyrics from popular songs of thirty years ago, and other ranges filled with the commands needed to use word processing software already a decade old and obsolete, nor yet again banks of memory taken up with a flood of paperback fiction and nonfiction read on trains, in bed, and on idle Sunday afternoons. If you have read many fewer words in your life, and perhaps read those fewer words over and over again, surely it is easier to remember more of them.

Again, it bears mention that in Jerome's environment comparative philological study had to be done relying chiefly on the memory. Lexica, indices, and encyclopedias were not at hand. This lack increased both the anxiety and the attentiveness with which he would read—once read, those words would disappear and be inaccessible except for what he remembered. How unlike ourselves, idly turning to the index at the back of the book to find something we read twenty minutes ago. Further, he needed fear no competition from more efficient technologies. Even if I thought that I knew the facts of usage of a rare word, I would never dare speak them out as he did, for I would be quite sure that some person with a far less retentive memory than I would loom up moments later, brandishing a dictionary or running a word-search on a computer, with a loud "Aha!" to show me that the word was used not three times but five or nine and in some really unexpected ways.

Other remarks on memory and its history will follow, but even to tell this small story about Jerome's memory is to engage in just the ahistorical juxtaposition that I have been identifying. We tell such stories in order to compare ourselves implicitly to Jerome, ignoring most of the differences and distances that separate us. If we could ever juxtapose the Jerome who wrote the letter identifying those three occurrences of a word with ourselves in any comprehensive way, we would be overwhelmed by the alienness of the man and never manage to see him as our rival or colleague. The portrait at the beginning of this introduction is already a thousand years anachronistic and idealized, while the fiercely ascetic Jerome of his Bethlehem hovel—it seems more familiar to call it a monastery, but what Jerome lived in was far poorer, shabbier, and smaller than the word "monastery" lets us imagine—would cut a most unattractive figure. (His famously complex relations with women remind us that his image is very much that of the scholar in *his* study. Not least of the effects of these historical reinscriptions is to reinforce expectations about all manner of social roles.)

One move that we could make in the name of history would be to jettison the past as irretrievable and irrelevant. When we complain that our contemporaries (it is always younger contemporaries we blame for this sort of sin) dismiss the past as irrelevant and betray an appalling—to older eyes—ignorance of history, it must be borne in mind that they thus have a defense firmly grounded in history itself. But we do not yet live wholly in a postmodern, posthistorical world, and I do not believe that we will ever quite jettison even the remote past. Too many of its physical remains are with us, and even as accurate knowledge of history is dismissed or neglected, our historical tourism and fetishization of monuments continues apace. We cannot recover our past or live it again, nor should we desire to do so; but we can learn from it, if we are cautious and patient and meticulous. For me the fruits of history are twofold. First, the pleasure of the act of "doing history" itself is considerable, enticement enough to pursue the pastime but not perhaps justification enough for the time passed on it. And second, the usefulness of history lies in the sharpening of sight, the heightened awareness of difference, the respect for nuance, and the sense of the possibilities of change.

To be sure, from Herodotus to our own day, the majority view of history is that it sets up models of virtue for emulation. Critical scholarship runs into a hail of rhetorical bullets when it tries to adjust the idealized past to conform to the actual surviving evidence. The most visible and influential professor of history in the United States in 1997, Speaker of the House of Representatives Newt Gingrich, chooses to take the easier route of idealization, at the expense of facts, in televised history lectures. My own allegiance to the harder-eyed school of professional history is based in part on my assessment of my own character. In history as in my own life, negative models ("I'll never treat people the way *he* does") are far more potent and durable forces in shaping my conduct for the good than paragons are, and less vulnerable to deflation.

What follow in the chapters of this book, therefore, are historical meditations that take as their point of departure the specific issue of how past western cultures have used the spoken and the written word as bearers of culture and as shapers of the world that human beings think they live in—what Rilke called "die gedeutete Welt," the interpreted world. They are studies of the various incarnate manifestations of that fundamental but slippery unit of discourse, the word. A word is a polymorphous thing, after all. I may think of it as a few sound waves echoing in an empty room as I muse aloud to myself, but it is almost certain that no one could think of a word as a discrete entity without the external visualization of writing to separate it from other entities like it. Words stand for things, but words are known to us only as signs for the things they thus create, or at least differentiate, by representing.

A word may indeed be a series of graphic symbols transmitted in any of a hundred ways (carved in stone, written with pen on paper, photocopied onto paper, or dancing on a screen in pixels), and those symbols may be alphabetic or pictographic (where the same image may find itself able to associate with several very different pronunciations). The function of such a graphic artifact is usually intermediary and always artificial, and in the study of that artificiality the best minds of every culture are driven near to distraction. Augustine spoke of words as "choice and precious vessels" for meaning, but immediately went on to rebuke the "wine of error" that they too often carry—as though words are merely instrumental. A modern sensibility will be more cautious about granting immunity to words, which can hold in themselves quite inextricably the failings of judgment and principle that beget them in specific forms.

For us, alive now in the last years of the twentieth century, no human culture is imaginable without words. We think of ourselves as language-making and language-using creatures, and we still take some pride in the distinction. But is this pride destined to be everlasting? Chia-Wei Woo, a distinguished physicist and president of Hong Kong University of Science and Technology, lately mused aloud of a time when language itself may seem to be an anachronism—when some postlinguistic means of communication renders the culture of the word obsolete. It is impossible to deny such a possibility, though it goes against current linguistic orthodoxy, but we should linger over the question whether any form of unmediated communication will ever be possible: After language, what? A metalanguage—that is to say, a system of surrogates for thought and reality that are more sophisticated, more supple, more effective than those we use as words?

We live in a historical moment when the media on which the word relies are changing their nature and extending their range to an extent not seen since the invention of movable type. The changes have been building through the twentieth century, as the spoken word reanimated communication over telephone and radio, and as the moving image on film and television supplemented the "mere" word. The invention and dissemination of the personal computer and now the explosive growth in links between those computers on the worldwide networks of the internet create a genuinely new and transformative environment. Zealots foolishly proclaim that the book is dead, and utopians and dystopians croon and keen over the futures their fantasies allow them. My own view is that we can expect no simple changes, that all changes will bring both costs and benefits, loss and gain, and that those of us fortunate enough to live in such exciting times will be put on our mettle to find ways to adapt technologies to our lives and our lives to technologies.

My purpose in writing this book has been to make it clearer what is happening or what might happen by thinking about similar transformations in the past, watching people's reactions to them, and then cautiously, thinking about the present and the future. The esthetic of closure and fixity that we now cherish may very well turn out to be one taste among many, and the possibility now coming into view of a world in which useful and persuasive discourse never has to choose to freeze itself, but can continue to grow, amend itself, ramify, and become more subtle and more true in response to its readers and to its author's continuing experience—that world may generate its own esthetic of openness and adaptability. As I wrote those words, some synapses begin to short out on me, for I know that the person I am today cannot grasp such lability—my words betray that, in that what I want to do is grasp and hold. But the world is a fluid place, and it does not always treat kindly those who would freeze it, or part of it, in place. Imagining our way forward to the people that we, or our children, may become is an exercise we need to undertake if we are to live through these changes successfully and make good use of the chances the world offers us.

I need to try to see myself not only in that painting of Jerome/Nicholas, but also in the fragments of a more nearly science-fictional future. To look back is reassuring, at least for those whose present is marked by relative privilege and comfort. It is comparably disconcerting to look ahead to a world in which the same standards of privilege and comfort may not apply. But those of us—even a culture's elite teachers, scientists, scholars, writers, artists, and intellectuals—who do not look forward in this way may expect to be marginalized in the struggles to make sense of ourselves and our world. My own choice, heart beating in my throat nervously as I make it, is to resist the temptation of the classicist's quiet study and to seek to understand the lines that lead from the ancient times I know professionally to my own time, and to try to trace where those lines may go beyond into the future. I study the past, but I plan to live in the future.

Niccolò Machiavelli was the most self-absorbed of men, but he was not mad. We acknowledge that instinctively in telling the most famous anecdote of his private life. He told a friend how, when cast out of office and living grumpily at his country house, he would finish the day's chores and then go to his writing room in the evening. There he would pause in a small entrance hallway to wash his hands and robe himself more formally, go into the writing room, and there hold conversation with the ancient worthies. It was in that room that he wrote some, at least, of the stream of books that would make him one of the first writers of best-sellers (and scandalous best-sellers at that) in the age of print. A middle-aged man who gets dressed up to go into his study by himself and talk to people who died centuries earlier is not always regarded as sane. But Machiavelli is an emblem of his time in an important way that also explains his behavior.

In English, it was Machiavelli's near-contemporary and (in a way) *Doppelgänger,* Thomas More, who first used the word "publish" to describe one of his literary activities. For though we speak of publication of ancient and medieval works, no such thing actually happened. Production of copies was done by handwriting, and even when we think we see booksellers producing multiple copies of the same author, the market scarcely ran beyond the ken of the author himself. The author of the manuscript book remained responsible for what he wrote, and the book itself was a personal production. There was no divorce between private and public.

But the print media effected that divorce. The act of writing was now "taken private," so to speak, while at the same time the act of making the results promiscuously public changed forever the conditions of discourse. Now you labored in silence and secrecy to perfect your words, then dispatched them to a world that knew you only by the book. The "author" was born.

Of course, this scheme exaggerates somewhat, and of course authorial qualities existed before. But the recognized role of the author and his separation from the words that went out to represent him to a wide public of strangers were phenomena that took on new intensity from the introduction of printing. Machiavelli in his study had found the privacy of authorship, but he could not yet quite cope with it. He was still too much the old man of face-to-face society, and so if he did not actually produce his works in a community of discourse, he counterfeited that community with his pretty and engaging story. Classicists have long quoted it fondly, thinking we see ourselves in it, the way we see ourselves in Jerome. It's another model, but a deceptive one. We already know how to be writers, we know how to be private, and we revel in that isolation from the world in the moment of creation, in order to revel in imposing ourselves on the world in the act of publication. (What do writers want when a book is published? Attention, acclaim, response, notoriety: they want the act of imposition to succeed in seizing the public stage, the stage that has been inaccessible until the act of publication occurs.) We are not Machiavelli, nor are we Jerome.

I now "publish" an online electronic book review journal. More than one of our reviewers has expressed surprise and delight at the experience of writing a review and dispatching it for publication, only to be able to see the publication almost instantaneously, and the reaction from the audience equally swift. And that is in a setting that enshrines the print model of submission, editing, and discrete publication. Elsewhere in cyberspace, far more of our private discourse reaches a wider public in near-instantaneous fashion than ever before, and sometimes we get flamed for our pains. But Machiavelli's interlocutors have disappeared from the room in which I work, and a world, literally, of friends, colleagues, and strangers, has succeeded them. Wherever I go, the community of friends and colleagues I "converse" with daily by email goes with me—even halfway around the globe. But when I say I publish my electronic journal, I am still calling on my experience in the old culture in exactly the way Machiavelli was when he went into his writing room. The challenge for us today, as it was for him, is to balance old models with new modes of behavior that exploit the possibilities of the new environment effectively without disorienting us so completely that we forget who we are. Machiavelli in dirty clothes, without the special writing room and the conversation of the worthies, wouldn't know how to be Machiavelli. Our own communities shape who we are no less forcefully, and our technologies give us power to shape those communities.

Chapter One

PHAEDRUS: HEARING SOCRATES, READING PLATO

At age 17, I wrote in an essay on a college application form that the two historical figures I would most like to have met were Socrates and Jesus. With any luck, I will never have to read that essay, but I remember the handwritten note on my letter of admission from the director of admissions saying that he wished he could be along when I met them. Because I didn't attend that college and can't be sure I wouldn't have had the experience of "meeting" Jesus and Socrates, my theme here perhaps should be regret, but instead I tell this tale to evoke its familiarity and triteness. The two names were calculated to please my elders and came to me from a tradition of middlebrow cultural studies. Sage moderns from Will Durant to Karl Jaspers would have known where I was coming from, and would have added the self-consciously wise observation that Jesus, Socrates, Confucius, and Buddha all represent comparable and (in the case of the last three) nearly contemporaneous manifestations of similar spirits of innovation and enlightenment in the major cultures of the ancient world. How odd and wonderful, the conventionalism goes, that such disparate cultures saw such similar visions. Was some benign ecumenical providence at work? If not, what explains this homology of sages across cultures over a relatively narrow window of time?

Consider some further points they have in common. First, none of the four figures was an unambiguous success in his lifetime. The prestige we assign to those four names comes from long after they lived. Each founder's school had its rivals, often for many centuries. Taoism stood soon enough in China against Confucius, and Mo Tsu was more outspoken in that direction. Buddhism swiftly divided into multiple sects, as did Christianity, even as Christianity also found itself rivaled from outside. In Greece multiple philosophical schools competed for the authentic heritage of philosophy, and Academic, Peripatetic, Stoic, and Epicurean became brand names for ideas. Of the four, only Confucius' followers flourished in his native land uninterrupted, and even they finally fell afoul of the enthusiasms of the late Maoist years. Buddhism survives and thrives away from the Indian homeland of Prince Siddhartha, where Hinduism rules. Christianity has had its moments in Palestine, but for most of the centuries since Christianity became a world religion, its homeland has been in non-Christian hands. Socrates left no cult, but the power of the Platonic tradition was at its greatest centuries after Socrates' lifetime in the rarefied form of neo-Platonism, and his home also fell under alien ideological control (first Christian, then Islamic) for many centuries.

Two other considerations will help clarify what I am getting at. First, all four of these men were received and revered as teachers: male authority figures, passing on their wisdom to young men in homosocial environments. These are not the heroes of war or commerce; rather they come from the less quantifiable world of ideas and values, where reputation depends heavily on reputation. Second, each of them is known to us from a written tradition (though none wrote books himself) which is unanimous in praise of its founder even while clearly containing material of highly various authenticity. (Nonlaudatory contemporary texts about these figures do not survive [Socrates is a partial exception], though one can see from the stories of their lives that none was uncontroverted or unopposed in his lifetime, and the difficulty pinning down the exact words and deeds of their founder is shared by all four schools.)

We need not see these men as extraordinary manifestations of charisma and wisdom. The real message of the apparent harmony of their teachings indeed lies in the platitudinous and benignly impractical nature of them. (For it is also the case that few of the followers of these leaders bears much resemblance in way of life or in wisdom of discourse to the founder.) They are instead better seen as comparable phenomena in two regards.

First, Jesus, Socrates, Confucius, and Buddha all come from a time in their native culture when several connected events were occurring. The introduction of literacy threatened the hegemony of traditional aristocracies, and at the same time increasing commerce and the connectedness of societies threatened the isolation and tranquility of single communities. Cities, with their nervously permeable boundaries and fluctuating populations, were an uncomfortable but essential part of the landscape within which the stories are set. Looked at in this way, each of the four, to say nothing of their contemporary and near contemporary competitors, idealized a countercultural set of ideas inimical to the hurly-burly of the thriving society of the time. Jesus represents Judean isolation and separatism at a point when a succession of empires had definitively brought Palestine into the wider world. Confucius could not find a prince to make him sage-in-chief and neither could

Socrates' disciple Plato—not for want of trying in either case. Socrates was famously out of temper with his home city of Athens. (In his case, his idealization of Athens misleads us—his simple city was dead and gone in the geopolitics of his time, destroyed by its own imperial folly. All these figures make hay out of giving voice to sentiments which many can approve and few espouse.

But there is a second feature they hold in common, and that is their reception in later centuries. Each has been successful in a line of tradition that can be traced with some difficulty from their time to the present, but their image depends heavily on having been gradually rewritten and, in the process, assimilated to western cultural norms. They fit our notion of what the countercultural sage of antiquity should have been like. Just as in their own societies they offered both a real alternative to a few misfits and an agreeable possibility to many more, so they represent to, let's say, a twentieth-century college admissions officer a piece of western culture or world civilization that justifies by its piety (and its impiety, since three of the four resisted the traditional religion of his society to some extent) the disorderly and often bloody chaos of a world that tries hard to live on platitudes.

But why should such figures emerge with particular clarity in societies in which the practices of literacy are becoming more widespread? What can we learn about the power of the word and its avatars? The case of Socrates is illuminating and gives westerners who examine it a sense of warm filial piety. (We think our young people should read Plato without asking what he might say or what effect he might have on them. His function is totemic, scarcely ideological.) Navigating the rights and wrongs of Socratic argument is good for orienting the mind to the mythic landscape of the West.

Socrates himself, of course, comes to us only in writing, and through several sources. Aristophanes' *Clouds* gives us a hearty comic take on the old gaffer written while he was still alive. He appears to be a bit daft but quite amusing. With great show of consideration, scholars linger over this image, rather glad to have it, but knowing that in the end they will reject it. More tellingly, after Socrates' execution in 399 B.C.E. Xenophon left memoirs of Socrates that give us a far less defined figure than that of Plato, and Xenophon was more or less Plato's exact contemporary. We know Plato's Socrates better than all these others. We enjoy penetrating behind the Platonic encrustations to find the real Socrates and to make him the gadfly, the anticlerical patron saint, that we need.

The final irony of Socrates is the one that his disciples, however, created for him. In making him the subject of texts, and then in making those texts the focus of a school (which quickly became just one of several competing schools), they succeeded in turning the dead Socrates into the one thing he had least time for in life: the figure of authority devoted to teaching wisdom.

That Socrates stands precisely on the boundary between the worlds of spoken and written discourse. The canonical texts themselves are dialogues, engaging transcripts of the interactive word. To find the heart of the ambiguity of Socrates, we must go to one of Plato's most charming and enigmatic creations, the *Phaedrus.* This little book tells a story of an afternoon's dalliance on a shady river bank between an old lover and his young paramour. There's no sex, because they're philosophers, but there might have been, and much of the discussion has direct relevance to the question of whether there would be. It is also a book about books, written about the subject of writing when the subject was still a relative novelty.

Here philosophical dialogue is brought before us as the alternative to sex. Without the *frisson* of wondering will they or won't they, the dialogue's opening speeches can seem bloodless and abstract. Instead they are witty deployments of the sexual and personal tension surrounding the relations between benevolent but sensual older men and lithe younger men in Athenian society. There was a bit of the forbidden about such encounters, but a little discreet sex play that let the older man have his fun was countenanced and probably even fashionable. The wit and more or less the wisdom of the *Phaedrus* lies in the way this fun culminates in a rapturous description of the powers of love, with no sex at all. Instead, the conclusion to which Socrates leads the discussion—after the false start of two first speeches offering a deliberately bloodless and cynical view of the relations between older and younger men—is that philosophical discourse, the rubbing together of minds through dialogue, is what leads to a spark of illumination and perception of fundamental truth: an orgasm proper to the intercourse of souls.

The *Phaedrus* is famous as a treatise on rhetoric, part of Socrates' running battle with the men he called sophists (and whom he resembled more closely than his disciples have been comfortable admitting). Do we imagine language as personal and persuasive, or as a common possession used responsibly to get at the objective truth? It would seem that the rhetorician (monologist and performer that he is) uses rhetoric as a tool to manipulate other people. Socrates offers a countermodel, whose differentiation he exaggerates, of interactive communication in dialogue. He speaks of "the serious treatment of these subjects [justice, for example] which you find when a man employs the art of dialectic, and, fastening upon a suitable soul, plants and sows in it truths accompanied by knowledge. Such truths can defend themselves as well as the man who planted them. They are not sterile, but contain a seed from which fresh truths spring up in other minds. In this way they secure immortality for it, and confer upon the man who possesses it the highest happiness which it is possible for a human being to enjoy" (276d–277a). In a work with similar themes, the *Seventh Letter* which is attributed to Plato and at least written in his spirit, the same dialectic process is described as rubbing ideas together and subjecting them to "tests in which questions and answers are exchanged in good faith and without malice so that finally, when human capacity is stretched to its limit, a spark of understanding and intelligence flashes out and illuminates the subject at issue" (344b).

The assumptions here merit pause. The relationship of speakers is still the authoritative older man as the teacher and the docile youth. The roles differ from that of the sophist teaching wisdom, or the rhetorician delivering speeches, to a mute audience (what Socrates disowns as "recitations that aim merely at creating 'belief' "). The difference lies in the way the ideas are received by the young man and by his active and flexible understanding and manipulation of them. The highest use of language is still unequal but collaborative, not a mutual seeking after truth but a form of indoctrination through dialogue. By the time Socrates is finished, he has effectively given an antimanual of rhetoric or a manual of antirhetoric, showing how language should be used to communicate truth that lies beyond language.

How well does Plato's representation of Socrates' dialogues demonstrate the process Socrates describes? A good case can be made that what we are meant to see in them is the process itself, even perhaps in the very last dialogues, where the Socrates figure rants on authoritatively but the form of dialogue is more or less maintained. What remains, however, is that we have all these written dialogues enshrining a man who expresses the deep skepticism of writing. Listen to him as he urges his hearer to believe "that a written composition on any subject must be to a large extent the creation of fancy; that nothing worth serious attention has ever been written in prose or verse" (277d). The parallel text of the *Seventh Letter* is equally dogmatic: "No serious student of serious things will make truth the helpless object of men's ill-will by committing it to writing . . . When one sees a written composition . . . one can be sure, if the writer is a serious man, that his book does not represent his most serious thoughts; they remain stored up in the noblest region of his personality. If he is really serious in what he has set down in writing 'then surely' not the gods but men 'have robbed him of his wits' "(344cd). (The quotation marks identify a line of Homer, probably quoted from memory rather than any written text.)

That skepticism is understood if we pay heed to the circumstances of reading and reception that Socrates imagines. The written word for him is mute, unable to explain itself. Written words, when queried, maintain the same "solemn silence" as a painting. If you ask them what they mean, they give the same answer over and over again, and they cannot distinguish between suitable and unsuitable readers. Writing has not the protections against misunderstanding that speech has.

Socrates' complaints were true and valid. Face-to-face conversation deservedly retains its prestige. But still we have accepted writing as a medium and now prefer it for just some of the reasons that Socrates finds speak against it. The objectivity and externality of writing is what we now value. Words that have been written down will lie still when you examine them—there are no speakers to change their minds or shift their ground. We have evolved techniques of analysis that can say to a fare-thee-well what the range of meanings in such words are. And when we examine the written word in that way, we and it achieve a functional equality that is lacking in Socrates. To be sure, our reverence for the author of written words is long implanted and strengthens the power of words, but we have learned to fight back in various ways. The reader today certainly does not imagine him- or herself primarily as the younger disciple of the author of a book or article, being schooled in the ways of the world. The egalitarianism we prize is something that Socrates could not imagine and would have found puzzling.

A key to understanding the Platonic dialogue form lies in what Socrates says late in the *Phaedrus* (276c), where he speaks of written exercises as a "pastime," the practice of amusing oneself with the composition of discourses. These words then fall on the last page of a text where Socrates expresses satisfaction at the discussion with which they have been amusing themselves. The dialogue as written is a model of how discourse might work, but it is no substitute for dialogue in the flesh. The long tradition that engages in substantial formal exegesis of Platonic texts is profoundly faithful to Plato's Socrates in every regard except the one that matters. It betrays its hero by taking itself seriously; by cutting itself off from dialogic pursuit of truth, it abandons the Socratic ideal.

How much of that ideal is worth retaining is a large question. We can cherish the dialogue and cherish face-to-face communication without dispraising other forms and other media. (If we do this wisely, we may see new media sometimes reenacting old forms with a new vitality.) The Socratic idealization of one form of communication is unrealistic in many ways. If we depend on face-to-face communication, then we condemn the larger social organizations on which we rely for sustenance. Communication beyond limitations of space and time—the great benefit of the written word—is denied us.

Recent interpreters have shown how Socrates' own language betrays the weakness of his idealization. Twice in the *Phaedrus* (276a and 278a) the authentic language of dialogue is spoken of by metaphor as "written on the soul of the hearer." Thus for Plato's Socrates, the written word is already so much a constitutive part of language that he cannot speak of or imagine the unwritten word except as not written, or, to use his metaphor, written in a different way, on that invisible other, the soul. The advantages Socrates imputes to dialogue emerge only when the comparison with writing is made. At the same time, no one would have invented writing without a clear sense of its advantages.

Socrates ascribes the invention of writing to a mythical Egyptian (274d), who claims that writing will be a drug that enhances memory and wisdom—a drug in our modern sense (as Derrida showed; we have been on Derridean ground for the last two paragraphs here), that is to say, of ambiguous power, one that may heal or poison. Socrates has the king of Egypt resist this claim, because Plato's Socrates has a famously idiosyncratic conception of memory deriving from his view of reincarnation—we learn nothing, he says, because we knew it all before, we must only be reminded of it. What the two views share, however, is a disinclination to examine the notion of memory, and even more the inability to distinguish versions of memory according to the expectations and experience of people who worry about it. Think of the anecdote about Jerome's memory: what memory can be asked to do changes as our cultural expectations change.

It is fairer to say that memory itself is culturally designed and redesigned over time. Greek and Roman writers speak with amazement of those who memorized the complete *Iliad* or *Aeneid,* as though such people retained some preliterate skill. But the very idea of spoken repetition of extensive texts is impossible to imagine in a nonliterate culture, for there is no sure way to test such a feat of memory. The bards of preliterate culture told the same story repeatedly, but with endless minor variation and (one trusts) liveliness. The mechanical recitation of texts depends on having a script to check, and no one can recall grade school poetry memorization exercises honestly except as voices droning more or less accurately on, quite immune to poetic inspiration, corrected only for inaccuracy.

Writing taught us a new model of memory. Our more recent ideas of memory are scarcely less fantastic, or less conditioned by technologies of information retrieval. To hear our contemporaries, one would think that memory is like an endless video- and audiotape record of all that we have experienced, and that it is only a largely inexplicable liveware error that makes it hard for us to retrieve the whole tape, but that if we try hard, we can and (the repressed-memory practitioners tell us) often do retrieve the whole tape. The ambiguities of those retrieved tapes, however, should remind us that the very idea of "memory" is an artificial one: that is, a faculty that records true things, things as they really are. Everything we know about psychology tells us that before the mind can record such things, the senses and the intelligence have edited and selected the realities in a way that introduces subjectivity and even error. But I would go further to suggest that the boundary in the mind between memory and fantasy is hardly watertight. Paradoxically, our anxieties about memory make us consciously try to distinguish the two. It is hard to tell (although perhaps anthropology can help us here) how far a culture without external records feels compelled to make human memory function as well as a document would. For the moment, my point is that Socrates' own category of "memory," implicit in the discussion between the inventor of writing and the king of Egypt, is already shaped by the fact of writing, to say nothing of Socrates' own idiosyncratic (and I believe largely reactive) ideas about how memory ought to work.

Plato's prestige is so great that many modern historians and theorists have been content to start with the axe-grinding philosopher in assessing the impact of writing and the rivalries it begot in ancient Greece. There has been lively debate in our time, largely fostered by Eric Havelock, the Torontonian classicist who later taught at Harvard and Yale. In the way of the 1950s and 1960s, Havelock insisted on drawing a hard line between "orality" and "textuality" and saw the Platonic moment we have been reliving as revealing the point when the "Greek mind" assimilated itself to the written word. In a schematic way, he makes many good points, and it is surely the case that much of what we cherish in the golden age of Greek literature stands on the borderline between oral tradition and the written word. The differences between Herodotus and Thucydides, for example, historian as storyteller and historian as investigator of "facts," are all too easily schematized as backward- and forward-looking.

More recent work has recovered both more of the sophistication of Herodotus, for example, and more of the traditionalism of Thucydides. Rosalind Thomas, in a well-grounded study of a wide range of the conditions of written and spoken words in classical Athens, is at pains to show what we should have expected: that oral and written practices existed side by side, often incompatible, but rarely examined or confronted. The great mistake is to imagine a sharp boundary created by a single development in society separating before and after. The lesson of historical investigation is that change brings complexity, and a suitable metaphor for social change will be multidimensional and disorienting—as disorienting a model as the real thing would be to have lived through.

The best perspective to put on Socratic ideas, the one that draws the right lessons from his skepticism, is the longer range history of Greek literate culture. The aristocratic and self-contained perfect city of Plato's fantasies (as in the *Republic*) did not come to pass. The *Seventh Letter* is our best source for how close Plato came in a failed alliance with a Sicilian prince, and its true lesson is that the philosopher found himself in practice unable to accept disciples on anything but his own extreme terms. His criticism of writing in that text begins paradoxically because a would-be disciple has written a book professing to embody Plato's doctrines, and it is the imperfect fidelity of the textual disciple that he rejects most strenuously. Plato's Sicilian adventure falls very far short of what the *Republic* imagines, and that is an important lesson.

More important, within a generation of Plato's lifetime, the physical and social conditions under which that fantasy could be indulged disappeared, as it turned out forever, in the Greek world. Alexander's conquests consolidated what had been plain to see (however often denied) for a century, that the Greek world could no longer consist of isolated communities linked by the occasional and formal bonds of colony and mother city. The "world" had been invented and could not be uninvented. Different communities now had to live with the knowledge that they were not autonomous or unique. Aristocrats would still cling to local authority, only to realize that merchants and soldiers showed a disconcerting ability to move around, acquire wealth, and even within a generation or so to become aristocrats themselves. The decisive refutation of Platonism is the historic fact of Hellenistic civilization—a civilization that venerated Plato and Socrates and began the process of idealizing the great teacher figure harking back to simpler times.

Socratic skepticism of the written word virtually disappeared, to be cited and mused upon occasionally, as the written word spread far and wide as a vehicle of business and culture. A truer representative of the Greek view of the written word, though itself a fantasy of a very different kind, is the totemic image—famed in our culture to this day and often invoked for an importance so self-evident that it cannot be true—of the library at Alexandria. Our legends of Western Civilization need Alexandria as the mother ship, as it were, of the western library tradition. But those legends as well need to deal with the fact that the Alexandrian library was destroyed, destroyed utterly. Not just Alexandria but the Hellenistic and Roman tradition of library-making went to nothing—texts survive, but no library of Greek and Roman antiquity survived. In an ancient context, that particular library enshrined a consciousness of community spread across time and space. To seek out all the books in the wider world showed a consciousness of that world and a sense of meaningful communication with it.

In the long run, the most important achievements of ancient imperialism had nothing to do with politics and everything to do with language. The spread of Greek throughout the eastern Mediterranean to become the lingua franca from the Balkans to the upper Nile was an extraordinary cultural achievement and created a long-lasting and powerful community. The later Roman achievement is even more important as the Romans not only spread their language over a still more vast landscape, but made it stick: Italy, France, Spain, and even Romania all still speak the language the Romans taught. We know too little of the oral side of the transmission of such languages, but what strikes the eye is that the educational and cultural institutions of Greece and Rome used the written word cannily to extend and maintain cultural hegemony.

Great cities and great empires were not the only communities that found ways to use the written word to assure their growth and expansion. I think it is not an accident that in the years since writing the college admissions essay, I have gone on reading some of the same books, but have found more in them than personal encounters with the wise of old. The multiple styles of reading that I now bring to those books—and exemplified in this chapter—are themselves heirs of a long process of invention and inheritance. We live in an age unprecedentedly fortunate in its recognition that reading is not one simple thing, but a related set of activities, each with its own power for enlightenment.

Chapter Two

FROM THE ALEXANDRIAN LIBRARY TO THE VIRTUAL LIBRARY AND BEYOND

The idea of the virtual library is a catchy and current idea: "California has stopped construction plans for new university libraries. Instead the state intends to focus its attention and budget on 'virtual libraries.' I gather that this means that all information will be available primarily via computer and CD ROM." (*Washington Post,* 21 November 1993). The virtual library is a dream that many share, something many have imagined but none has seen. The main feature of this vision is a vast (ideally universal) collection of information with instantaneous access to that information wherever it physically resides.

A search for the phrase in computerized databases (which themselves offer a hint of the imminent arrival of fast online resources) reveals that it is indeed a recent coinage, recent, at least, in anything like the current sense. People have long spoken of something that is virtually a library as a virtual library, but it was little more than a decade ago that computer journalists first bandied the idea about: "Consider, too, virtual libraries where you enter into a rich virtual stack space and 'browse' about." For that visionary, the idea involved special goggles and gloves, a feature that seems to have faded from most recent discussions of the virtual library. The phrase burst into the popular press only as recently as 1987, where the helmets are still in place to "give the student the ability to go inside a chip, visually. We can give him a tour through a virtual library or a virtual museum."

So the phrase suggests a vast collection instantly accessed . . . But phrased in those terms, the idea is much older. Few recent discussions fail to mention an article by Vannevar Bush that appeared in the *Atlantic Monthly* in 1945, imagining a device he called the "Memex," but they could also cite the article in the *New York Times* of September 10, 1950, which described the "Doken," a high-speed reading machine which could search the contents of all the books in the Library of Congress in ten seconds. Note that the function of the new and amazingly fast (even by 1997 standards) technology is presented as a tool for using a traditional, familiar, famously all-possessing institutional library of the present. The social and intellectual structures of the present are assumed to be stable and to be a useful basis for understanding a new technology.

Although the fantasy has been around now for at least two generations, we still think of it as a modern one that looks beyond the book itself to a visionary's future. Starting from that assumption, this chapter began not as a eulogy for the virtual library's past, but as a hymn of praise for its revolutionary future. I even had a perfect visual resource to use as a foil to this future virtual library, a monument to the old library.

Twenty years ago, at a festival of the films of Alain Resnais, I saw a short documentary he made in 1956 called "Toute la mémoire du monde," on the ways and working of the Bibliothèque Nationale (BN) in Paris. I remembered it as a period piece, singing the praises of a grand institution just as it entered the autumn of its existence. Finding the film again took some trouble, for it was apparently never released in this country, and when I did, it took me (thick-headedly) some time to realize that what I was seeing was very unlike what I had remembered. To be sure, it was a film that praised every aspect of the BN, but the terms in which it did so were, on a repeat viewing, suddenly revelatory.

Those who know Resnais's early work would recognize the dark and lugubrious tone of the film, but would be amused by its depiction of life in the library. Everything is mechanized. The row of postmen bringing sacks of books marches stiffly to circuslike music, and then a huge apparatus of indexers, cataloguers, fiche-makers, spine-stampers, and shelvers mark the progress of the book through a librarian's assembly line, until with swift precision the book reaches the waiting hands in the great reading room. To anyone who knows the BN, of course, that swift, unfailing service is just what marks this film out as a fantasy, instead of a documentary. But this time it was the substance of the vision that finally hit home. The huge collection, containing "all" the world's memory, was in its time already the virtual library we now think lies just ahead. Imagine the narrator's voice from the film for a moment: "In an instant, it [the new book] becomes part of a universal memory, abstract, indifferent, where all the books are equal among themselves, where they all enjoy together an attention as tenderly aloof as that of God for man. And here it is chosen, preferred, made indispensable to its reader, pulled from its galaxy."

To see this dream of a universal library in a different technological setting from our own is not merely to recognize that the dream existed then, but that the existing and foreseeable technology of 1956 looked like a fully satisfactory way of achieving it. Today's dream of a universal library is weighed down with silicon chips, keyboards, screens, headsets, and other cumbersome equipment, but someday another dream will make today's imaginings seem as outmoded as the fiche-makers and shelvers.

What persists, has persisted, and I think will persist, is the combination of ambition and self-deception in the ideas that people have in common about their present state of affairs. What was excellent in the BN of 1956 was that it already was the thing that we now think we must wait another decade or two to recapture. What has changed is not the dream, but the sense of technical possibilities.

The dream itself of a virtual library is much older than 1956, and its history can help make clear why it still has power. The notion of a library is an extraordinary one, of course, and thus fragile. Surely it is not self-evident that the words of other times and places, frozen forever in unchanging form, should live on indefinitely, in ever accumulating geometrically expanding heaps; even less self-evident that human beings preoccupied with the real problems of their present should spend any appreciable amount of time decoding and interpreting the frozen words written by people long dead. That it seems self-evident to do so says something important about the culture that was created using writing and print, but also says indirectly that this culture is contingent, malleable, and far from any final form of human organization of knowledge.

If the essential feature of the idea of the virtual library is the combination of total inclusiveness and near-instantaneous access, then the fantasy is almost coterminous with the history of the book itself. The earliest invocation of such a dream is a famous document of the second century B.C.E., the "Letter of Aristeas to Philocrates." The letter introduces and justifies the existence of the first major Greek translation of the Hebrew scriptures, the so-called Septuagint; in it also, the author attributes to Demetrius of Phalerum (the "minister of culture" for Hellenistic Egyptian king Ptolemy, the founder of the "library" at Alexandria a century or so earlier) the ambition of gathering together, if possible, all the books in the world. Even the pigeonholes into which the papyrus scrolls at Alexandria were to have fit play their part in the virtual library legend: this economical and easily managed form of packaging would speed the access readers wanted.

The library at Alexandria has long loomed as a chimera of power and mystery on the horizon of our culture, but the real makings of our tradition are less ancient than that and clearly betray again the presence of the fantasy of the virtual library. That tradition is Latin in its origins, European in its development, and now western in its self-presentation to the world. There are many reasons for what happened in the Latin tradition, some of which have to do with the cultural transformation traditionally discussed as the "fall of the Roman empire," but the phenomena were complex and marked by innovations whose effects have lasted into the present. One decisive event was the introduction of the codex for formal literary use, something that happened between the second and fifth centuries C.E. This meant that the books prepared and used in the old way, on papyrus rolls, were obsolete and few would survive in the Latin west. Texts from those rolls that survived did so by being copied over into the codex form of bound pages (usually on animal skins rather than papyrus).

The discontinuity between ancient and late antique library communities overlaps and closely (but not exclusively) resembles a discontinuity between the traditional literary culture of antiquity and the chiefly monastic Christian textual culture of the middle ages. Further, the discontinuity that emerged between Greek and Latin is important.

Classical Latin literature always lived under the shadow of Greek literature, but in late antiquity the Greek shadow passed and Latin began to live on its own. Christian Latin literature in particular sensed its obligation to its Greek past, but had little or no linguistic aptitude for confronting or cherishing that past. From roughly the fifth century C.E. the western Mediterranean and its dependencies to the north and west became wholly, independently Latin, save where the Romance and Germanic vernaculars began to come into use. In that fifth century a self-conscious tradition emerged that took written texts seriously and began to organize them theoretically in a way that was embodied from place to place in real collections of books.

Compare for example two influential books of that period giving advice to the Christian scholar. Augustine wrote *On Christian Doctrine* around the year 397, setting out principles for the interpretation of scripture. Itself a sign of the text-centered nature of the religion he represented, that treatise is short on bibliographic pointers to any secondary literature. Augustine writes as a bishop and therefore (by virtue of his ordination) an authority. He has no need to quote the specialist literature. A hundred and fifty years later, Cassiodorus, a retired statesman living at the monastery he had founded on his family's estate on the remote southern shore of Italy, wrote *Institutes* for the student of scripture. Now it was the private scholar's turn to provide this guidance, and when he did so, his work was little more than an annotated bibliography, arranged by books of the Bible, of all that had been thought and said in Latin (or translated into Latin from the Greek, and Cassiodorus had a staff of translators working on adding to that literature-in-translation) that could help the Christian exegete. To this was added a further bibliography of specialist literature in related disciplines, including the so-called liberal arts, for Cassiodorus believed that rhetoric and astronomy were useful things for the Bible student to know.

In the years that had intervened between Augustine and Cassiodorus, the Latin Christian community had learned to depend on texts for many things. Augustine himself had kept a catalog of his own writings and even wrote a treatise late in life listing and defending all the books he had written; he was no longer just a charismatic preacher, but a man who had a written record to defend. In the same period, Jerome's translation of the Bible (which would come to be called the Vulgate) had begun to standardize the text of scripture in common use. The library of exegesis had boomed in size. Augustine's own oeuvre, for example, amounted to over five million words, so much that it was another sixth-century monk, Eugippius of Naples, who put together the first instantiation of "The Essential Augustine" (an anthology about one thousand printed pages long) to help the reader who wanted swift access to the saint's writings.

About the same time, Dionysius Exiguus, a Scythian (more or less equivalent to Ukrainian, at least for geography) monk, found that the Latin churches were often embarrassed by a lack of accurate information about church law and so put together the first collection of what would later be called canon law. This change can also be marked by comparison with Augustine's life: In the late 390s, Augustine in north Africa had run afoul of church law enacted at the council of Nicea seventy years earlier by being ordained bishop while his predecessor was still alive, but he had the very good excuse that his church did not have a copy of the relevant decisions and so did not know what the law was.

Also in this period the bishops of Rome began to claim the exclusive title of *papa* (pope) and to use the written word in such forms as chancery documents to advance their authority. Thus we have not only large collections of papal letters casting their influence over a swath of Europe from Yorkshire to Constantinople, but also form letters, that is, evidence of a chancery so busy that it had verbal templates to use. From a slightly later period there comes something called the "Daily Book" *(Liber Diurnus)* of the popes, containing just such a collection of form letters. Power sprang from the pages of a book. That kind of power was increasingly centralized and the autonomy of the local community weakened.

Similar phenomena appear in this period across a wide spectrum. Take this charming story, for example. One Sunday morning in the fifth century (according to a historian writing about a hundred years later, so perhaps this is best thought of as a story about the world ca. 590 C.E.), bishop Sidonius Apollinaris of Clermont in southern France was going into church to conduct the eucharistic service, when a prankster snatched from his hand the pamphlet *(libellus)* that had written in it the prayers he would use at that service. The point of the story was that not only did the bishop carry off the service from memory with distinction, but the congregation was surprised and delighted that he did so—people now expected him to be dependent on a text.

As it happens, Sidonius probably wrote what was in that pamphlet himself, but his is the transitional age in which such things were being written down and propagated. Within a few decades of the telling of that story, it would be the norm in a city's large church on Sunday for there to be no fewer than four large books in use at once to guide the service: a sacramentary for the bishop to follow, a lectionary from which a deacon would do scriptural readings, an *ordo* in the hands of the master of ceremonies making sure that people didn't bump into each other, and a gradual (music book) in the hands of the choirmaster.

No longer was this the spontaneous early Christian community in which the spirit-blessed spoke freely what they knew to be the truth. Words could now be chosen and crafted in advance, not even necessarily by those who would say them. Liturgical participants, even the bishop, were actors in a scripted drama—a significant change. The process begun there would culminate a thousand years later at the Council of Trent in the publication of a missal book for use in churches that would prescribe to the priest not only every word, but every gesture, and often even which fingers to use for each gesture of the service. By that time, a large part of the power and the authority of the liturgy had left the church building and gone into centralized hands. Control over texts had brought control over people.

The pattern is familiar. Such centralization is of course both costly and beneficial. What is lost in autonomy and spontaneity is gained (we like to think) in assurance, control, consistency, and predictability. My point is not to decry the development but to pinpoint an important stage in its advancement. It was in the fourth through sixth centuries in the Latin west that our cultural ancestors created a particular set of software (chancery letters and missal books, for example) with which to manage their lives. Taken to its most extreme form, this led to small communities organized around a kind of text that Jesus never imagined, the monastic rule. Something like the familiar *Rule of Saint Benedict* or the less well-known and more obsessively orderly *Rule of the Master* can suggest what it was like to live in such a community. Yet the most important feature of such a text is not what it does but *that* it does what it does: it makes the life of a community depend neither on spontaneous choice nor on the orally assimilated customs and wisdom of the past nor again on a charismatic leader, but rather on specific rules and regulations written down on a page. The Benedictine rule insists that it be itself read out to every novice several times, and then read out again in pieces to the whole monastic community repeatedly in an endless cycle of renewal of textual authority.

The reliance on texts I have just documented implies that someone will own texts and they will be accessible. Late antique library collections are not very well documented, and that gap in the evidence is regretful. There were libraries at Rome under papal supervision, and the monastic collections of Eugippius and Cassiodorus left interesting traces. There is even a vignette from the same Sidonius describing his own library, with separate seating for men and women: the women decorously surrounded by the works of the church fathers, the men at the end of the room where the dangerous "pagans" were shelved.

There exist today collections of manuscripts that have lived together under constant care since the fifth century C.E. in the Latin world. Verona is one place to find a few such manuscripts, the Vatican Library the best place. But our real knowledge of libraries begins at a later period, the time of the so-called Carolingian Renaissance of the ninth century, when the many libraries that emerged are known to be direct heirs of the late antique collections. The selection and ordering of texts in these libraries closely follows the principles of the fifth and sixth centuries. Works like Augustine's autobiographical listing of his own works were responsible in many places for libraries' holding a high percentage of those works. These medieval collections exhibit a modern, not an ancient, arrangement, placing objective truth at the center of the collection and organizing everything else around it not for beauty but for utility. In this period, what we know consider the classics were found well down the list in a library, if at all, and subjoined to the basic texts of Latin grammar they presumably exemplified.

The history of books and libraries in the central and later middle ages is abundantly documented and well and widely studied. In his signal new interpretation, *The Implications of Literacy* (Princeton, 1982), Brian Stock reveals the richness implicit in realizing that there were many overlapping textual communities shaped around virtual libraries—that is, collections of texts shared in spirit and in fact by groups of readers and their associates.

The invention of printing changed many things, and that story has also been told often. But some essential things did not change. Despite massive disruptions, the fundamental community of producers and users of texts remained fairly constant—the same clergy and aristocrats. Some ex-monks turned into university professors to be sure (Luther is a prime example), but the continuity of the community of texts was in the main intensified. The codex remained the outward form of the book and the techniques that exploited its power in the late years of an exclusively manuscript culture were enhanced rather than supplanted. Indexes, cross-references, and tabulations all multiplied. Many libraries that flourished in the later middle ages still survive today, often at the heart of still vital institutions, even physically, as in the case of Duke Humphrey's room in Oxford's Bodleian Library. Where late antiquity had seen disruption and the creation of a new tradition, early modernity had every opportunity and reason to transform its inheritance, but instead turned remarkably conservative in the face of the possibility of chaos. The deliberate emphasis on and systematic reacquisition of Greek and Latin classical literature created the illusion of a tradition stretching back well beyond the late antique origins of the library tradition and incorporating Greek and Roman antiquity in a single traditional continuum.

But it is not just the temporal dimension that is important. Historically, cultures dependent on the written word have all shared the fantasy of the virtual library—that is, they have cherished some notion of total inclusiveness. What they achieve is always far short of anything that might be considered a totality of output of the written word for even a brief period (even the great depository libraries contain only a fraction of the printed reading matter of their own societies), and they have placed a high value on access to that totality. But with this vision, physical institutions have grown up that in one way or another impersonate the virtual library. When I haunted the small public libraries of my childhood, they already embodied for me, imperfectly but inspiringly, the virtual library I had in my imagination. The research libraries I prowl today are equally coherent and equally tantalizing in the way they suggest even greater riches lying beyond them. In both cases the fantasy is far from reality, but potent nevertheless.

The fantasy that a library's users share defines the community to which they belong. It embodies a worldview (thus the "nonfiction" worldview of the virtual library of the age of the codex differed from the "poetic" worldview of antiquity's collections) and so seems to give objective confirmation to what we believe. What we mean by a comprehensive collection of books, for example, is exactly dependent on who we are. In that way the imagined library functions importantly as a transmitter of culture from one generation to the next, as it did to me in those public libraries of my childhood.

What of the tradition of the virtual library will survive in the coming networked communities? The written word itself will surely see its grasp weaken, as it dances in tandem with visual and aural treasures in great abundance. Other familiar landmarks will diminish.

The author is already an endangered species, and rightly so. The notion that authoritative discourse comes with a single monologic voice thrives on the written artifact. Both oral discourse (before and beyond the written word) and the networked conversations that already surround us suggest that in the dialogue of conflicting voices, a fuller representation of the world may be found. The notion that reality itself can be reduced to a single model universally shared is at best a useful fiction, at worst a hallucination that will turn out to have been dependent on the written word for its ubiquity and power.

Similarly, the notion that discourse must be fixed to be valid will fade. Fixity is to our eyes the only satisfactory guarantee of authenticity (the U.S. Copyright Act requires fixity as a condition of its protection), but fixity brings with it rapid obsolescence. There is scarcely a page I have published in a decade and a half of scholarly writing that I would not now change if I could, but I cannot. Words that I know to be inadequate and in some cases untrue continue to speak for me. I am no longer the person I was when I wrote them, but I am still somehow their author.

With the idea of fixity goes the idea of duration. Good words are thought to be words that last and remain unchanged. But if the world is constantly in flux, then surely the descriptions of that world should find a way to change to reflect that changed world. Some of our reference works do this already; Jane Austen, however, is perhaps immune to rewriting and should remain so. (I say this even while noting that adding to the canon of dead authors is a recent fad that suggests that some of the intellectual implications of the changes of which I speak are beginning to be felt, to the detriment of the reputations of authors from Jane Austen to Margaret Mitchell.)

The greatest transformation that such an environment will bring is in the way culture is transmitted. If the idea of a stable, reassuring set of texts and truths on which to nourish the young fades, then it will not be at all clear what it is we need to do to or with our young people to acculturate them to the ways of their elders. For years I have quoted with amusement the poet John Crowe Ransom in an essay on Princeton, in which he concludes that all in all Princeton is a fine place, but if he had a son, he would just as soon lock him in a library until he was twenty-one, then send him to Paris. The value of Paris at age twenty-one remains about what it was, but even in jest the notion of locking a youth in a library will, I suspect, so far lose its power that people will not understand what Ransom was getting at. The idea that the totality of our culture can in some way be incorporated in a library is what will disappear. In that way, the virtual library reveals itself as a very conservative dream.

If the virtual library is fifteen hundred years old, why does it still seem so current? Will it continue to enthrall? I am fond of quoting media guru Marshall McLuhan's notion that the content of a new medium of communication is always imagined to be another older medium. Thus cinema at the outset was thought to be a vehicle for filming plays, and there are still "made-for-TV movies" and "TV newsmagazines." A farmer at the turn of the century saw that the horseless carriage could get him to market and back more quickly, but had no inkling that the same vehicle would send an interstate highway through his pasture and change his way of life forever. It takes several generations to get past the point of depending on the old medium for a way to think about the new and to get to the point of exploiting the new medium artfully in its own right.

The dream of the virtual library comes forward now not because it promises an exciting future, but because it promises a future that will be just like the past, only better and faster. No one can deny the usefulness of such conceptions, but their limitations must be recognized as well. The virtual library may already be obsolete.

The vital difference between present and future practices will be that the forms of organization of knowledge in electronic media do not resemble those of the traditional codex book. The methods of production and distribution will diverge from those of the print media even more. Where the library has traditionally been one of a few such enterprises cooperating (if sometimes at arm's length) with a finite community of publishers (and thus both together functioning as gatekeepers on a limited set of narrow information pathways from authors to readers), a community is now growing in which there will be as many publishers as readers. The possibility even of imagining totality in such a world rapidly disintegrates. What would be the contents of the electronic virtual library? Everything? Every what? Just to ask the question makes it suddenly obvious that one of the most valuable functions of the traditional library has been not its inclusivity but its exclusivity, its discerning judgment that keeps out as many things as it keeps in. We have grown up assuming that information is a scarce resource and devised our economics accordingly; but in an information waterfall, the virtual library that tells us everything and sweeps us off our feet with a cataract of data will not be highly prized. The librarian will have to be a more active participant in staving off "infochaos." If the traditional librarian has been conceived as a figure at home in the discreet silences and cautious dealings of a Henry James novel, now perhaps the right model will be found in James Fenimore Cooper or the Star Wars films: something between the pathfinder Natty Bumppo and the Jedi knight.

Hyperlink

THE INSTABILITY OF THE TEXT

The culture of print has inculcated the expectation that a given author's words may be frozen once and for all into a fixed and lasting pattern, one that readers can depend on finding whenever they find a copy of a particular book. Broadly speaking, that expectation is valid, but even in the world of print there are reservations to be had. It is surprising what variations can occur between one printed edition of the same book and another, and if the work is a classic, often printed by different houses over a long period of time, a burgeoning of variant readings can arise comparable to those in works copied in manuscript. How much effort it is worth to identify and delimit such abundance is debatable. A generation ago there was such a debate between the Modern Language Association, on the one hand, approving a series of new editions of standard texts brought to a high level of editorial accuracy, and the critic Edmund Wilson on the other hand, who found in the project a superabundance of pedantic detail and little benefit.

But no one would disagree that before the relative stability of printing, texts were often disconcertingly labile and unreliable. No modern edition of a classic widely copied in the middle ages has been possible without an attempt to construct the family tree of resemblances and disresemblances among the manuscripts that survive, and then to find a path back to the remotest ancestor we can reconstruct. There lies one caution for readers of ancient books. To see a cleanly edited, intelligently punctuated text of Cicero is to get a very inaccurate impression of how these words would have appeared in antiquity. Our standards of the orderly page, marked by consistent visual punctuation and clear wordbreaks, simply did not obtain. By comparison to the neatly weeded and pruned gardens of words in which we take our literary pleasures, the ancients made do in a wilderness of irregular scratches on a page, and made do quite well.

A linear progression from chaos to order cannot be imposed on the history of the presentation of the word, though, for we have now returned to a time of instability marked by debate over means of presentation. To enter a text in a computer means to make choices. It is possible to make the simplest possible set of choices and to allow the text to take the form of a series of Roman alphabet characters, upper and lower case, delimited by carriage returns, tabs, and a handful of standard punctuation marks. For many purposes such a text may suffice. But it must be borne in mind that this is already a highly encoded form of representation, depending on a recognized convention that joins together what the computer reads as eight separate digits (1s or 0s) into a unit (or byte) and allowing 128 such combinations to stand for numbers, letters, and punctuation marks. Texts represented with these 128 characters are widely transportable from one computer system to another. If we knew how to be content with simple texts, this might be the basis of a universal orderly representation of texts.

But already 128 characters are insufficient to render a text in, say, French or German, where characters require accents and umlauts. A first attempt to render this environment more flexible, introduced by IBM at the time of the first personal computers, has doubled the character set to 256, including a fairly wide set of marked vowels for some European languages (but notably omitting Scandinavian requirements) and a few Greek letters of use to scientists. (The added characters are so variably and eccentrically chosen that the story is told, perhaps an urban legend, of their being selected by two IBM employees on an overnight flight to Europe with a deadline for a morning meeting. If that's not how they were chosen, it may as well have been.) Virtually every computer in use has embedded in itself awareness of the conventional 256-character set of signs known as ASCII (for American Standard Code for Information Interchange).

But this way madness lies. A new universal alphabet is in the making, to replace the 256 characters now known to computers with a set running to about 35,000, embracing every distinct symbol in every writing system known to humankind. The Unicode project creating this alphabet has been a distinct intellectual adventure in its own right: When do two symbols merely provide variant ways of writing the same sign? When do two symbols used in different languages and looking superficially alike really differ and when may they be counted as a single symbol? Once coded into every computer, Unicode would allow a far more flexible representation of all the writing systems of the world. But there are already too many computers in use in the world to promise rapid replacement of what we have.

Letters in a sequence are only part of a text. The struggles of users of early word processing software to get their expensive new machines to do what they had been able to do with typewriters were a reminder of the importance of layout and typography in the communication of information. Typesetters of the last century, before heavy mechanization took over, estimated that they spent only a third to a half of their time putting characters in sequence, and the rest of their time arranging the white space on the page that surrounded the characters. We now have a myriad of ways to introduce formatting information into our computerized texts, and the results are beautiful to see on screen or on paper. But there is a cost in the loss of standardization. An abundance of word processing formats has generated another abundance of would-be standard formats. Recognition of these formats depends on users' choices of hardware and software. If, for example, I need to get tax forms from the U.S. Treasury, I can find them on the World Wide Web and print them at home in minutes. But I must have previously acquired one of (at last count) four different ways to manage text (PDF, PCL, PostScript, or SGML) in order to get those forms at all. Each of those names represents a very different conception of how text may be arranged in a computer.

Roughly, such schemes divide between those that seek to describe the visual arrangement of a text that might be printed and those that seek to describe the structure of the information that goes into a document. PDF, for example, stands of portable document format, and depends on software created and distributed by the Adobe Corporation. It allows readers with the right software to receive files easily and display them with all the typographic features of print on their screens. But such text is hard to edit and search because the graphic page representation inserts coding that gets in the way of the original string of characters. At the opposite extreme, a text in SGML (standard generalized markup language) is presented only in the ubiquitously available ASCII characters and is described in a way that any computer anywhere may read. But an SGML text per se is still a mass of text interspersed with codes that require special software to be read intelligibly. (The HTML on which the World Wide Web relies is a simpler version of SGML.)

The result of this disagreement about ways of representing electronic texts is that any given collection of texts in a computer will be a mix of usually incompatible kinds of codes requiring different software and often different hardware to interpret. Some textual communities will use one or another coding system fairly consistently (scientists are fond of a language called TeX, for example, but few humanists or commercial users have ever seen it), but we are very far from any possibility of mutual interchangeability.

The obvious inconvenience of this state of affairs—the difficulty of managing a "library" of such texts—masks a more troubling problem. Computers, and their software, change rapidly. Bodies of information created for a computer are marked by the kind of coding possible and necessary at the time of their creation. As the environment changes, it is usually necessary to make at least some small changes in the text to keep it readable. If the format is particularly idiosyncratic (say, if the text has been coded by a CD-ROM manufacturer determined to present text in a proprietary way) or simply if several generations of transformations of software and hardware intervene, a text can become impossible to read. (I still own a box of 5 1/4" diskettes and could probably find a place to read them, but how much longer will that be true? CD-ROMs may last as long as a few decades.) NASA has found this problem particularly troubling: reels and reels of tape bearing computer data from the 1960s are now, at best, a series of 1s and 0s, while the hardware and software that created them have long since been rendered obsolete and destroyed. It takes a "data archaeologist" to attempt to decipher what has been lost.

Give us another generation of proliferation and surely vast quantities of information will slip away from us this way. We will no longer be able to depend on survival of information through benign neglect. There are medieval manuscript books that may have lain unread for hundreds of years, but offered their treasures to the first reader who found and tried them. An electronic text subjected to the same degree of neglect is unlikely to survive five years.

None of this is good news for librarians. Whereas the codex book in print form has been a remarkably standardized and stable medium, subject mainly to the depredations of material aging (crumbling paper and breaking binding), the new flood of electronic texts brings with it an exponential increase in the difficulties of making information available to users and preserving it over time. Reader demand will be only moderately helpful in determining how society's institutions act, for as readers we will want both the newest and the best in everything, and permanence and reliability as well. We can't have both.

Chapter Three

FROM THE CODEX PAGE TO THE HOMEPAGE

Among the Mediterranean peoples whom we acknowledge as ancestors, writing began on stone, but became a useful vehicle for the dissemination of information only when the technology of the papyrus roll, or scroll, was perfected. Though the Athens of Sophocles and Plato (ca. 400 B.C.E.) was a town almost intoxicated with the power of the written word, the real achievement of ancient literacy was consolidated more than a century later in the Greek-founded Egyptian city of Alexandria. There the rows on rows of neatly docketed rolls in their pigeonholes created both the first great library and the first generations of literary critics and their readers. Most of the treasures of that library are now lost to us, and what survives are, for the most part, fragments unearthed in the last century from the Egyptian sands.

The ancient papyrus roll was elegant to look at but cumbersome to use. Size was strictly limited. We are told, perhaps only in legend but in verisimilar legend, that the length of a "book" in antiquity (such as the twenty-four books that make up the *Iliad* or the *Odyssey*) was de facto defined by the size of the pigeonholes at Alexandria. A thousand or so lines of text was all that a roll could hold, and that would already make a long sheet of papyrus, averaging twenty to thirty feet in length. To shuffle through such a roll looking for a passage was time-consuming and bothersome. To manipulate twenty-four of those in order to control one of Homer's epics would be to enter a seriously user-hostile environment.

There was already a humbler form of information-processing technology at hand for day-to-day purposes: the wax tablet. In its simplest form, this was a slab of wood, hollowed out in the center and filled with melted wax. A dry stylus would incise letters which the thumb could then erase at will. These tablets were excellent for memoranda, transient bookkeeping, and daily business. Several of them bound together by thongs made something roughly like a small notebook.

Somewhere in the first centuries of the common era, the notion of making a formal literary medium out of bound pages on the model of the wax tablet suddenly began to catch on. Though the codex—for such is the name given to the form in antiquity—could have pages of either papyrus or animal skins, in practice parchment and vellum (made from the skins of sheep and calves, respectively) began to be most widely used. Use of this codex form for literary texts became appreciable in the second century, during the third century the corner was turned, and by the fourth century, the codex had won the day: of surviving second-century Greek manuscripts, 99 percent are rolls; of surviving fifth-century manuscripts, 90 percent are codices.

In any case, the manuscripts of the Christian Bible are predominantly transmitted to us in the codex form, and there is much discussion about why. Some propose that the codex form was chosen for its ease of reference and cross-reference. That is hard to substantiate, since none of the traces of such reference systems can be found in those earliest manuscripts. Even the notion of a Bible as such was slow to form. Scripture was transmitted for the most part in separate manuscripts containing parts of the whole. The Greeks had larger comprehensive volumes earlier, but the Latins waited very late, probably until the sixth century, before drawing all the books of their scripture together in one set of covers.

One implication of this change must be emphasized. If you were a very farsighted text of the second century and you wanted to be read a thousand or more years later, the thing you most wanted was to be copied into a codex format. Books that made that transition successfully had a reasonable chance of surviving and being read in the centuries to come, while books that did not were more likely to be orphaned. The Greek originals of the plays of Menander, for example, which were not copied into codex form, were almost entirely lost to us until modern discoveries in Egypt began to restore them. The Latin plays of Plautus and Terence, on the other hand, some of them not much more than ham-handed translations from the Greek, had a long and lively history of medieval and modern readership. They had been rescued from the roll and saved for the codex at an early and timely date.

Transference did not guarantee survival, of course but it was a necessary condition. It is worth bearing in mind that when we think of scanning our libraries into machine-readable form, we are making a similar judgment about their fates. Put another way, too much attention to preservation of the printed book may have the perverse effect of undermining prospects of future readership if materials fail to be digitized. (That judgment is rendered increasingly draconian by the decay of much twentieth-century archival and library material preserved on acid-based paper. It simply cannot be economical to preserve indefinitely every library's every copy of every book, and the sheer quantity of discards that will be forced by decay over the next few decades will at first astonish us.)

For a variety of reasons, we in the western world are the heirs of Greece chiefly at one remove, through Roman and Latin hands. The living tradition of Greek culture in the middle ages withdrew (from our point of view) to the Aegean shores and the Bosporus, a veil of Islamic and Slavic cultures fell between east and west and closed the Mediterranean to the free commerce and intercourse of antiquity, and western civilization was left to develop north of the Mediterranean and west of the Vistula.

Within those realms, a remarkable culture came into existence. St. Augustine (354–430 C.E.) stands at the head of the line of that new culture. When Augustine set out on a literary career that left five million words still surviving today, there was little substantial Christian literature—in role or codex form—for him to work with outside the Bible. Within his own generation, the first Christian bibliographic literature was being written, and within a century, library management was a new and pressing topic for many Latin writers. Augustine appeared to his community as a mediator of the written word contained in the large and handsome Gospel book that stood in his church sanctuary, a book that inspired his own production in great abundance.

Cassiodorus, in contrast, is more fully a man of this new culture. He is associated with the most ambitious single project we know of from his time, to create a Christian university at Rome, and after that a Christian library on his estates near Squillace on the Ionian Sea. He commissioned a portrait of the Hebrew scribe Ezra (who restored the books of the Law after the Babylonian captivity) for a Bible manuscript in his library, and the portrait of Ezra was probably at least implicitly a flattering portrait of the scholar and librarian in question—the bookshelves in the background are arranged according to Cassiodorus's description for his own library. By this time, the arrangement of bound codex volumes in an *armarium* (where they usually lay flat on the shelves, with titles on their spines), was a de facto standard.

The codex had several advantages over the roll. First, since its size was limited only by the strength of the user (or the user's furniture), much more material could be contained in a single unit. Second, the codex could be taken apart, put together, and rearranged at will. This meant that several different authors and titles could be combined and recombined with minimal difficulty. Third, and of greatest importance, nonlinear access to the material in the volume was possible. By this I mean simply that readers did not need to shuffle through every page from beginning to end to find quickly what they sought. With appropriate indexing or dumb luck they could pop the book open in the middle and quickly find what they were looking for. That third feature offers the genesis of the revolution in how words could be used for which the codex stood and which, in turn, offers the most important key to my reflections about our present situation.

Following again McLuhan's rule that the content of a new medium is always an old medium, the manuscript was first conceived to be no more than a prompt-script for the spoken word, a place to look to find out what to say. The arrangement of words on the page, without punctuation or word division, was as user-hostile as DOS could ever hope to be, and was meant for the technician, who knew how to use it to produce the audible word.

In many ways, then, the introduction of writing did little to change the way people used words, at least at first. And so there grew up a whole literature and culture of memory aids. Early medieval manuscripts feature diagrams accompanied by illustrations that offered mnemonic keys to scheme outlining, say, the nine kinds of syllogism or the twelve types of definition possible in the school taxonomies of the time. The people who produced those manuscripts seem still to have expected readers to manage their own private nonlinear information access system, which we call memory. The real advantage of the written word—that the nonlinear system it offers is one that anyone can use and that the user needs to know very little to get a great deal out of—had not yet been fully exploited. In the age of memory, in order to know something, it was necessary to know it, or to know personally someone who already knew it. In the age of writing, in contrast, it is possible to know things without committing them to memory, and that is a very great revolution indeed.

But the history of medieval manuscripts is the history of the exploitation of the possibilities of the codex page. Arrangements of material on the page made information more accessible and facilitated cross-movements of various kinds. The simplest example is the alphabet, and increasing facility with alphabetization, although it developed with agonizing slowness, is an important part of the history.

Consider, for example, the "Eusebian canon tables." One of the commonest early medieval Christian books is the Gospel book. One of its most obvious features is that it contains four narratives of the same life (of Jesus), all different. Much modern ink is spilled on the "synoptic question" (how Matthew, Mark, and Luke are related, and why John seems so unrelated to the other three), but the topic is not new. In the second century the Christian writer Tatian had already produced a work he called the *Diatessaron* (roughly *Four-in-One*) in which he reduced the four accounts to a single linear narrative. The idea was attractive but it did not catch on. In the early fifth century,

Augustine wrote at length on "The Agreement of the Evangelists," by which he meant their essential agreement in spite of the appearance of disagreements.

Medieval Gospel manuscripts themselves often include a simple reader's help. Each Gospel is marked with a running series of marginal numbers in sequence, starting over at the number one with each Gospel. Then in the front of the manuscript are pages in which architectural ornament highlights columns of parallel numbers. The technique is to locate passages in which, say, Matthew, Mark, and Luke all tell the same story. The marginal numbers from Matthew appear in the first column, with those of the corresponding elements in the corresponding stories from Mark and Luke in columns two and three. There are as many sets of these parallel columns as there are possible combinations of stories, so that there is a separate comparison for stories in Mark and Luke but not Matthew, and so forth.

An arrangement like this—standard in early Gospel books and traditionally attributed to Eusebius of Caesarea, the Greek church historian of the fourth century—is a form of nonlinear access for both the linear reader of the text (who reads a story in Luke and cannot quite remember where the parallels are in Matthew and Mark) and for the student who wishes to analyze overall patterns of coincidence and opposition. Such an arrangement of information at the front of the handsome Gospel book suggests a style of reading that does not simply start on page one and work through to the end.

The resourcefulness and the range of possibilities suggested by these examples demonstrate that the codex page format lent itself admirably to nonlinear access. The reader can now begin to fill in many of the other ways in which these techniques have advanced and refined themselves since that time. The index, the concordance, the page number, and the running head: all of these have medieval antecedents and modern application. Ever since the fourth century, we have been at work in Latin and post-Latin culture building a common tree of knowledge, an invisible but powerful structure by which we agree together to organize what we know and to make it accessible. The institutions we call libraries and the catalogs they contain (in whatever form, from old bound books through the familiar cards and down to our online library systems of today) are all only manifestations of a larger cultural project: to make knowledge available to nonlinear access in as many ways as possible.

If we reflect for a moment, we will find that a very high percentage of the forms of the printed word we consult most often are not linear narratives: travel guides, textbooks, *The World Almanac,* encyclopedias, dictionaries, phone books, the *Physicians' Desk Reference,* cookbooks, atlases, *Books in Print,* or *Ulrich's International Periodical Directory.* Consider as well the case of the learned journal, whose collocation of articles represents both a coincidence of subject at some level and at the same time a diversity of specific topics that often produces a quite dissonant effect. We know how to read all those nonlinear, atonal publications quite harmoniously.

The learned journal offers a prism through which to see the future of non-linear reading. It is one of the artifacts of non-linearity and it belongs properly to the history of the printed rather than the written word, a history that, if considered under two aspects, can help us catch sight of its and our future.

The traditional scholarly journal offers a material convenience. Binding together small items makes for an ease of access that is limited to the technology of the print world. The distinction between the article and the monograph is artificial—it is crude but effective to say that an article is a monograph that is too short to be printed and bound separately. In this form the journal is a relic of the age of print.

But the journal is also part of the larger cultural project. Whereas the earliest journals had a unity of place and time (that is, they represented a wide range of observations about the world of learning reported to and from a particular location, say Paris or London, at the date of issue), the modern journal has gradually been specialized by its place on the invisible tree of knowledge. I read the *Journal of Roman Studies* or *Behavioral and Brain Sciences* because I know that in those pages I will find a range of related studies that appeal to me. One could imagine that assigning each article a Library of Congress (LC) catalog number would improve access (and short of that, we surround ourselves with indexes of various kinds), but it must be admitted that, as a first cut on sorting information, when I read *Behavioral and Brain Sciences,* I am mightily relieved not to find studies of the philosophical inconsistencies of Seneca or the poetics of self-pity in Ovid in those pages, just as I am happy not to find B. F. Skinner's ideas extensively ventilated in the *Journal of Roman Studies.* While not without its drawbacks, this sorting function of the scholarly journal is one of its most valuable, and this is a function which is not tied to the technology used to create and distribute issues of the printed journal.

The reigning monarch of scholarly publication, however—the eminent monograph from a distinguished press—is in serious jeopardy. The traditional monograph, with its sustained linear argument, its extraordinarily high costs of publication and distribution, and its numerous inefficiencies of access, is beginning to look more and more like a great lumbering dinosaur.

This is not to say that the traditional scholarly book will disappear overnight, but surely its presence will fade. It will survive rather the way the leather bound edition of the classics now survives, not so much to be read but to make a statement about the book and about the owner of the book. For a time, academics will continue to delegate a large part of the responsibility for tenure decisions to university press editorial boards. In time, even that will pass; new techniques for arbitrariness and avoidance of responsibility will emerge, and young scholars will no longer speak with such misplaced reverence and awe of the publishing process.

Other aspects of our learned publishing will fade soon as well. Those aspects of the scholarly journal that depend on the physical media of printing and binding are evanescent. The issue, which for convenience yokes together forever items whose only commonality is usually their date of submission, need not any longer govern association. We will no longer refer to published material by codex page number, and so will surrender a remarkably imprecise and error-prone form of indexing. Better news is still to come. Electronic storage of and access to learned journals is a principal preoccupation of librarians and publishers. At the moment one touchy point is quantitative: how many copies of an article will be printed, and when? Publishers would like them all printed at once and sold to subscribers. Users have no objection to seeing fewer printed and sold and more copies photo- or electro-reproduced and distributed by interlibrary loan and other document delivery systems. That is a controversy whose life will be short: initial printing will disappear and any production of hard copy that occurs will be incidental and for the convenience of single end-users only.

It is easy to prognosticate that changes will come to the division of labor among the many existing journals. The proliferation of journals reflects both the crowded submission lists of the most popular ones and an urge to provide more specialized information through more narrowly focused collections. Both of these pressures can be met more cheerfully in a postprint world. A popular journal need not, in principle, limit itself by cost and number of pages any longer, but can instead use the time and talents of its editors as the measure of its capacity: how much can *they* stand to read?

Similarly, the subject association of articles is no longer going to be determined forever by their place of publication. No longer will it be necessary to agonize over whether to put an article in the *Journal of Roman Studies* or the *Journal of Hellenic Studies;* there will be no real reason why both labels could not appear on a single article. For the likeliest development (and here I am merely guessing) is that the association of articles in a given journal will no longer be a physical association and no longer be a condition for publication and distribution. The rest of the world will soon follow the scientific practice of publishing by distribution of something like preprints. The journal's peer review and stamp of approval will come after the fact of distribution and will exist as a way of helping identify high-quality work and work of interest to specific audiences. In that world, the journal title will be something like a *Good Housekeeping* seal of approval, applied after the fact; and there will be no reason, intellectual or economic, to deny a single article as many different such seals as editorial boards see fit. Indeed, we can already see articles that would reach different audiences if only they could be published twice in different places. We frown on this because trees and ink and shelf space are scarce resources. But if instead of multiple publication it were a matter merely of multiple electronic tags, then the form of indexing and access-enhancement that comes from identifying an article with the approval given it by a specific editorial board could be made much more valuable.

The first obvious consequence of these changes is that the quality of our nonlinear access to information will increase exponentially. To give only a hint of this, consider the difference between using Library of Congress subject headings for a subject search in an electronic catalog and pursuit of the same inquiry using keywords. In a world in which the library will cease to be a warehouse and become instead a software system, the value of the institution will lie in the sophistication, versatility, and power of its indexing and searching capacities. We need not wait for the possibilities of artificial intelligence to manifest themselves in order to take advantage of intellectually simpler but nonetheless powerful systems of investigation that can lead us through a mass of material to information that suits our needs.

The secret is that the end-user's intelligence remains a powerful tool. If a system leads me only close to the information I am looking for, I will recognize it and begin processing it in ways that no machine could. Artificial intelligence someday may begin to do some of the processing that I would do, but as long as my own intelligence is available, I need much less than a machine does to make sense of my world.

But one corollary of that style of searching is that I will not care where I find information, so long as I find it. A graphical display of the arrangement of information on a hard disk reveals that a computer is far less fastidious than a librarian. The information that seems to be stored so carefully is mashed together in a mighty jumble, pieces of files interleaved with pieces of other files, and bits of deleted files (not really deleted but merely unlabeled) strewn between. In that world, preservation of the boundaries separating one package of information from another is necessary only if the end-user needs it, and search strategies that concentrate on the information rather than the source are far more efficient and will thus be more successful.

One way to describe this phenomenon is to say that hyperlinks between data will become the dominant lines of travel from one item to another. But that is already true now and was already true in the medieval codex. The user would move from one manuscript to another, and from the indexing canon tables at the front of the Gospel book to one Gospel passage and then to another. The computer simply makes it possible to do more of that, and faster.

Another way of describing this phenomenon is to say that the boundaries that separate one information source from another are of variable value. Users for centuries have been ignoring them when convenient—consider yourself trying to remember whether you read that book review in the *New York Times,* the *New York Review of Books,* or *TLS.* When I have a hundred books on my shelf and I want to know when Cosmas Indicopleustes wrote his marvelous treatise in defense of a biblical doctrine of a flat earth, I little care where I find the information until I have to come back and cite it—that is, provide a form of non-linear access for others. The learned writer knows how hard it can be to track down again an item found once before among his or her own books.

Other boundaries will blur as well. For example, information that is gathered collectively, over time, with minimal consultation and organization but with equal zeal and care by people who have never met each other, may make up large and important databases. Here we will encounter what may be the fundamental conflict of interest in scholarly publishing: that between the freedom to speak one's mind and the responsibility to produce information that is assuredly valid and reusable by others. Freedom of inquiry and speech demand a world in which all may speak their minds, but the need for quality control demands a world in which we give power to people who are editors when we like them and censors when we do not. However that tension works itself out, an important but flawed or preliminary treatment of some vital subject will, by the time it has been worked over, discussed, revised, enhanced, and reworked by as many hands as care to turn to the job will become the ultimate postmodern authorless creation. Keeping the many cooks from spoiling the broth will be important, but bringing together all the world's available talent to solve a given problem will be a luxury we rarely have today.

A world in which it is not quite clear who is the author of a collective, cumulative, and collaborative work of scholarship may sound very novel, but it is also very old. The late middle ages had already created such books, like the famous *Glossa Ordinaria,* the common and widely disseminated medieval Bible commentary whose origins are still shrouded in mystery and which continued to grow and be relevant for centuries. We have come since that time to relish and rely upon the fixity of printed information, but that fixity will soon seem vulnerable. I have every confidence in our collective neurotic ability to cling to the value of that fixity, but at the same time I look forward eagerly to the flexibility and vitality of a medium that, as it plunges forward to the technical cutting edge, still pursues the enthusiasms and uses the techniques of the medium that it leaves behind.

Hyperlink

THE SHRINE OF NONLINEAR READING

Much has been written and much said about the library past and future. I want here to juxtapose two images that can give some idea whither we come and whither we go.

I should say first that libraries are the great love objects of my life. I remember a narrow Quonset hut with a single aisle of children's books at an army post in the western desert with deep affection, and the next somewhat more permanent building they built there, where I looked in vain for years for the biography of Mickey Mantle they were supposed to have but must have lost (to my delight I found an autographed copy for twenty-five cents at a used book sale twenty years later). We moved to the big city—feebly supplied with books and libraries in those days—but there was a branch library we could walk to, a Carnegie library downtown, and a decent little library at my high school. I suppose I could even say that my first job (volunteer) was typing catalog cards in that high school library.

But it was going to college and finding myself suddenly granted the run of the open stacks of a great research library that well-nigh swamped my consciousness. The riches of that place and its successors during my graduate school years were an ocean that I did my best to drain dry. Since then, I have become agreeably amphibian, drifting in and out of library buildings with armloads of books. I have at this writing five valid library cards and somehow there is always room to bring a few more books home, but harder to find room in my briefcase to bring them back—a curious gravitational effect. But what do I find there? Just a lot of books?

Consider for a moment a shelf in any library you might visit. What strikes you first is the diversity of materials there. The commonality of the codex form connects physical artifacts that may have come from all over the world and range in age from a few weeks to a few centuries. There is an anglophone bias in our libraries in the United States, but even so the books you handle may come from England or America, or possibly Canada or Australia, and will have been printed typically any time in the last century. But this shelf may also have books written and printed in at least a handful of European countries and, depending on subject matter, other continents as well. (I emphasize here the physical origin of the artifacts: there is another cultural marvel in the business of translating and republishing the world's intellectual product, but that act of integration takes place outside the library proper.)

These materials are not haphazardly placed, but rigorously organized, according to cataloging systems of great beauty. The utility of each book is significantly enhanced by that organized placement side by side with the others. For instance, even if you had a library where books were all shelved on some other principle, say by date of acquisition, you would need to know how to find them through some subject-oriented cataloging system to make them useful at all. The limiting factor, to which we will return, is that each book must choose one place to reside, though there may be multiple subjects in a single volume. (As we just saw, the scholarly journal suffers from this particularly: the library's catalog does not index individual articles, and so many subjects are clumped together under single titles and single call numbers.)

The library not only shapes but also creates the resources we see. No one could depend on bookstores for all that we get from libraries. Libraries are the after-market stock managers of the world's publishers. When the publishers have wearied of a book and made all they can from it, the librarian takes it over, cherishes it, rebinds it, lets you read it, chides you to bring it back when you forget, and eventually worries about photocopying it or otherwise preserving it when it grows old and tired. Scholarly research would be crippled without libraries.

One other feature of the library shelf needs mention: it is in constant flux. Most American libraries allow books to circulate, and so there are always books that aren't actually there, but are known to belong there and can be recalled if necessary. The structure of the library's stacks is similarly open—new volumes are always being inserted at their own point in the run of shelves, and the whole collection gradually balloons.

The collection is also chronologically diverse. We are familiar with looking at shelves of books published over decades. I recently had the experience of seeing the shelves in a brand new university where all the books had been bought within the last five years. There was something almost frighteningly clean and regular about the shelves of shiny new bindings, all uniformly tagged. It felt more like a bookstore than a library.

This thought experiment highlights the flexibility, the diversity, and the subtlety of the machine we have constructed. It also invokes a picture not of the future—for I would not pretend to know what the library of the future will look like—but of the present. How has the library changed already?

Nostalgia about our old card catalogs can easily blur the real issues surrounding their eclipse. First, given the cost savings of electronic technology, it would simply be irresponsible for a major library to maintain a traditional card catalog. Second, the power of the new technology vastly transcends the old. But third, the integration of the new with the old already changes the nature of the collection it describes and expands its uses.

The catalog of my university's library is now accessible from any continent on earth through the Internet, and I consult it regularly when I am far from home: that is already a change. But also, this catalog is a tool that describes the library's collection—by virtue of its technology—far more flexibly than the card catalog ever could; and that enriches the collection it describes, even if the collection itself is not changed.

For example, in 1973, in graduate school, I spent half an hour one day trying to find a specific author in the old card catalog. He was the medieval Irish philosopher known variously as John the Scot or as Erigena. I went on a chase through cross-reference cards in file drawers: I bounced from Scotus to Scottus to John to Johannes to Erigena to the variant spelling Eriugena. I found individual items, but I never found the author file for this author, and gradually I began to feel like Pooh and Piglet on the trail of the woozle, going round and round the bush, the tracks getting more numerous as more woozles joined the parade, but no woozle appeared. Finally I stopped, took a deep breath, and went on the chase one more time to discover that the cards had indeed been correctly directing me to the drawer with the Latin form of the writer's first name—but the spelling used was crucially "Joannes" not "Johannes." I had seen this half a dozen times at least on my paper chase, but my eye had not registered it and I kept going to the drawer for "Johannes." With a guffaw, I went to the requisite drawer and went to work.

To compare my memory with the present reality, I performed this same search on our online catalog a few minutes ago. It took three steps and less than a minute to get to the full set of records I sought. The path I took was one of literally dozens that, depending on what I had asked first, would have taken me to the same goal just as efficiently. (When I went back not long ago to the same card catalog in which my original odyssey played out—still open to the public but with no new additions since 1977—I found that my experience had been negated by the instability of catalog subject headings. All the "Joannes" cards have now been moved over to the "Erigena" drawer, a transition made at about the time scholars were firmly coming around to agree that this was *not* the correct way to spell his name.)

Now when I get to that virtual shelf we were imagining earlier, I have gotten someplace I could never go before. All the library's holdings on my Irish philosopher are displayed before me, including (1) those in the main collection on the Library of Congress shelves, (2) those in special collections around the campus, (3) those in the rare book collection under special care, and (4) those not actually in the building just now but out on loan. (Those in the main collection with the older books on Dewey Decimal shelves are a few keystrokes away.) It can tell me where each book is and when the books that are checked out will be due. Our library has a prototype of what some other places have done more ambitiously, namely a system for placing requests for recalls and even, on some campuses, delivery to an office. To be sure, it would be a pleasure to have the full text of all the books available online behind their bibliographic record, but that is a fantasy for now. My point is that the collection is already a different thing because of new ways I can know and use it.

It is better too for questions I can ask that I could not ask before. Keyword searches let me search for topics of my own devising, without depending on a librarian to have selected it as a proper subject heading, and let me combine terms to refine searches. All the traditional subject headings are there, so if I do not know what I am looking for, I may try my own combination, then inspect a record to see what other ways of asking the question might be there. If at first a search is fruitless, I can submit another within seconds, and I can roam about the electronic catalog with far more flexibility than among rows and rows of catalog drawers. In those days, when I made my way through the rows and found myself in the Ws, a query that would take me back to the Bs might very well go unpursued. Now it too is a few keystrokes away.

Finally, even today our libraries are pointing beyond their physical collections in creative ways. Our collection allows those of us with University of Pennsylvania IDs to search the following resources: Medline (online medical journal abstracts—a resource famously interesting to amateurs and hypochondriacs as much as to physicians); RLIN (a union catalog of research libraries)—if my library does not have the book, I can consult this resource to see where it exists and use the information to order an interlibrary loan; *MLA Bibliography* (a huge resource of literary scholarship, including journal articles, indexed in great detail); the *Oxford English Dictionary* online (I gave away my old two-volume compact edition a couple of years ago: I hadn't consulted it in years); Lexis-Nexis (online newspapers, licensed to let us consult it for classroom use); Dow Jones News/Retrieval; *Books in Print;* several more specialized indexes of scholarly and scientific journal literature, and then a world beyond all that as well, through the library's pointers to Internet information sources publicly accessible. (My own university's catalog on the Internet is joined by hundreds of others. There is real value to being able to consult, for example, the catalogs of European libraries from the comfort of my study in the United States to search for publications not readily accessible in this country.)

To describe how the Internet has already become a kind of alternate library would require another volume, and dozens of versions of that volume already exist, becoming obsolete hourly. But in one important regard, the Internet is *not* a library: nobody built it. There is great value in the diversity and abundance of information out there, and one may reasonably expect that diversity and abundance to continue to explode. But the qualities that make the library valuable are not quite there yet.

There is no organized cataloging, there is no commitment to preservation, there is no support system to help you find the difficult or missing resource. Finally, there is no filter: that is, there is none of the sense that a user of a great library has that somebody has thought about the possibilities and selected a set of materials to be both comprehensive and yet delimited. On the Internet, you never know what you're missing. That may change. Or we may change. For the moment, the library is still the most powerful paradigm for the organization and management of knowledge ever invented.

Chapter Four

THE PERSISTENCE OF THE OLD AND THE PRAGMATICS OF THE NEW

First a little late news from the Net. Books are here to stay: BOOKS ARE FOREVER, SAYS AUTHOR: Fiction Pulitzer Prize winner E. Annie Proulx says that the information highway is "for bulletin boards on esoteric subjects, reference works, lists and news—timely, utilitarian information, efficiently pulled through the wires. Nobody is going to sit down and read a novel on a twitchy little screen. Ever" (*New York Times,* 5/26/94, A13). Of course, a successful novelist has a certain conflict of interest that might distort judgment. Whether the fictions that we can and will read on screen will be known as novels is a question I leave for others, but the status of the book is surely as labile now as it has not been in five hundred years. To see what became of the technology, its critics, and the social order at the time of the introduction of print to Europe, and to see this not only as a triumphant but as a complex tale of fractioning and regrouping, puts us in a much better position to see the way forward from here in our own time of transition.

Let me begin with a story that played itself out on the strand at Rimini, then as now an Italian resort town on the Adriatic: Pope Symmachus was not happy. He had been summoned from Rome to the king's court at Ravenna to face accusations, not for the first time, of irregular election and unseemly conduct. To be sure, he had not yet been called, as he would later be, to pay back a rather substantial loan made from a Milanese banker taken to sustain the costs of doing business with the grasping functionaries of the government; but on the other hand, he had not yet been raised, as he would also later be, to the formal rank of Saint of the universal church.

He was this day breaking his journey to Ravenna at Rimini. As he walked on the beach taking the air, he saw a carriage pass by on the high road to Ravenna carrying women he recognized. They were the very women with whom he had been accused at Rome of illicit association, and he knew at once that they had been summoned to Ravenna to testify against him. Keeping this knowledge to himself, he waited until nightfall, took a single aide, and fled back to Rome, walling himself up in St. Peter's to hold out against the world.

The important thing about this story is the way we come to know it. Sometime in the early sixth century, it began to be the practice at Rome for a sketch of the life of each succeeding pope to be added to a collection of short biographies of his predecessors in what came to be known as the *Liber Pontificalis* (Priestly Book). The orthodox collection of these lives was continued well into the middle ages and is a standard historical source, whose value for information about popes, particularly from the late fifth century onward, is well known. But this story does not come from the orthodox collection, the one that helped make Symmachus a saint. Instead, it is found on the first three folios of a sixth-century manuscript preserved today in Verona. This unofficial version of the life of Symmachus, written according to the rules of the genre embodied in the official *Liber Pontificalis,* is followed by short entries, also resembling in form the larger *Liber,* about the other popes down to Pope Vigilius, whose death around 555 C.E. gives the manuscript a likely approximate date.

The fifth and sixth centuries, we have seen, were a period in which Latin Christianity was making remarkable strides in adapting to its use the power of the written word. It was not that Latin Christians were beginning to write, but that they were now using the written word with sophistication to organize and control their world. The papacy in particular emerges in this period as a textualized artifact. The lives of the popes are one piece of evidence, their chancery-collected letters that begin to survive in substantial numbers are another. The affair in which Symmachus found himself embroiled, the so-called Laurentian schism (named after his rival), testifies to the new power of the written word as well. First, the very survival of this "Laurentian fragment" suggests that not only was the papacy manufacturing itself by publishing official lives, but it was doing so well enough that it made it worth somebody else's while to write the counterhistory, to put together a collection in which Symmachus appeared, but not in a good light.

The text with which I began was probably written in ill-grace by the losers after the fact. An even more interesting collection of texts came from the winners in the course of the action. The so-called Symmachan Apocrypha are four texts written at the height of the controversy between Symmachus and Laurentius. They purport to be historical documents—a papal decree, the records of a synod, an account of a trial—dating from as far back as two hundred years earlier, all bearing in one way or another on the history of the papacy. What they have in common is that they provide elegantly forged precedents for various points of argument that Symmachus was making against his rival. The coherence and consistency of the implicit case these four texts make, and the very different faces they are made to wear, encourage us to say that these were not idle cases of stories told to good effect, but a deliberate creation of textual authorities whose authors had to know they were cooking (up) the books. It must have been worth forging those texts, and there must have been an audience in place ready to judge ecclesiastical legitimacy in documentary or textual terms.

Much else from that period shows both how the "library of the Fathers" was coming into existence with all manner of supporting documentation, and how far the public and spiritual life of Christianity was beginning to be regulated by authoritative texts. Vincent of Lérins, author in the 430s of a *Commonitorium* that attacked (discreetly and anonymously) the authority of Augustine of Hippo, may himself have been the first author to speak of the "fathers" of the church as figures of textual authority. In one sense this was inevitable, for the generation of Ambrose, Jerome, and Augustine had a very thin collection of Latin texts on which to rely and could not depend on textual authority. Those writers and their contemporaries left behind, by comparison, a vast body of work of imposing power and persuasiveness, and the next generations who read their books had to begin to make sense of what it was like to live in a Christianity where the bishop in your church's pulpit was surrounded by a throng of dead writers. By the sixth century the power of dead writers was felt strongly enough that it began to make sense to condemn them. A vital issue at and before the second Council of Constantinople of 553 was whether authors who had died in the peace of the church could be subject to retrospective condemnation. In one sense it was outrageous that such postponed judgment should be imposed. But if the question is not the living deeds, but the dead textual word of an author that lives on after him, then condemning authors was only an acknowledgment that books had begun to dominate discourse.

Although the late antique Latin experience in the making and shaping of power and community through the written word had a technological basis in the adoption of the codex form of the book, the real change was cultural and social. So too, at other moments of transformation, the original technological impulse is channeled by thousands of choices made by individuals and institutions into the shaping of society and its institutions.

The introduction of movable type made a revolution, no question. That story has been often and vividly told. It bears remark that the story of that revolution is regularly told in print and by partisans of the revolution. It may very well be that this revolution was a good thing, but any historical event recounted entirely by partisans is open to reconsideration. What happens when we ask those partisan narrators questions they are not very well equipped to answer: Who didn't like the technology of print, and why didn't they like it?

The applicability of these questions to our own time is obvious. If we can return to the last comparable watershed between ways of recording and distributing words—the introduction of movable type—and look for the history of resistance to the new technology in that period, we can gain some perspective on the controversies in our own time, when it is far from clear to many people that the revolution that is upon us will be a benign one. But how to find the answer to that question is not so obvious. The occasional mentions that are made in our histories of resistance to print are for the most part inaccurate, and thoroughly patronizing. But as we persist, we find more than we expect.

A good place to begin is with a standard popularization of the history of print, S. H. Steinberg's *Five Hundred Years of Printing,* published forty years ago in a Penguin edition. There we read of another great man who also passed time in the vicinity of Rimini: Duke Federigo of Urbino, the legendary bookman and father of the duke *fainéant* who figures in Castiglione's *The Book of the Courtier.* In the memoirs of his agent Vespasiano da Bisticci (ca. 1490) we are told that in his library "all books were superlatively good and written with the pen; had there been one printed book, it would have been ashamed in such company." The argument is from esthetic rather than utilitarian grounds, and the Duke must be given full credit: the Urbino manuscripts now in the Vatican Library give ample evidence of an elegance and an artistry of presentation that few printed books have ever rivaled. The Urbino Bible, for example, is a massive triumph of elaborate manuscript illumination. A theologian may well prefer a more workaday copy of a philologically sound critical edition of the scriptural text, and may even prefer a text in some language other than Latin, but few would deny that the devout are at least as well served by the grandeur of this book as by the shabbiness of modern puritanical black binding and fine print. But Federigo also had numerous lesser works copied into manuscript from printed sources.

In the same history we read of Cardinal Giuliano da Rovere, later Pope Julius II, patron of Michelangelo and father of the modern St. Peter's Basilica, who had Appian's *Civil Wars* copied in 1479 from the printed Speier edition of 1472. He retained the colophon and changed its wording from "impressit Vendelinus" ("Wendelin printed") to "scripsit Franciscus Tianus" ("Franciscus Tianus wrote/copied").

It may come as a surprise to those who do not work in this period that people were having printed books copied by hand. The authoritative study of the topic is that of M. D. Reeve, who makes several points. First, printing so multiplied the number of copies available that someone looking for a book to copy was more likely to hit on a printed than a handwritten exemplar. Further, if, say, the *Consolatio ad Liviam* (a fifteenth-century discovery attributed to Ovid, since recognized as spurious) is found printed in the works of Ovid, many who wanted just that poem would have copied it out, rather than buying the whole book. For these first two reasons, at least 10 of 16 and possibly 12 of 16 manuscripts of the *Consolatio ad Liviam* derive from printed editions. Third, fifteenth-century readers did not automatically prefer printed editions on textual grounds; in fact, they had reason to be suspicious of them. Hence printed books were not chosen as models, or at least cannot be proved to have been chosen, on those grounds. Finally, a few idiosyncratic owners account for most of the manuscripts made from printed books. Apart from Duke Federigo, the two bibliophiles Bartolomeo Fonzio and Raphael de Marcatellis were important patrons of hand-copied books.

The latter of these patrons, Raphael de Marcatellis, abbot at Ghent who died in 1508, has been the object of two important studies by Albert Derolez. These studies make clear that two considerations were foremost in the minds of a patron intent on a large collection of hand-copied books. First, luxury—and a preference for luxury is often a preference for an older, handmade product without any prejudice to the general utility or inevitability of mass-produced goods—and second, access, for hand copying was also a way of getting books not yet circulating in their own country. In the case of Marcatellis, more than half the incunabula (books printed before 1501) that served as models for his manuscripts are not to be found in any present-day Belgian collection. They were probably not on the book market in the Low

Countries, so he would have borrowed them from Italian friends. Hand copying was also the only way to obtain a coherent corpus of works on a given subject. We now praise electronic texts for ease of copying and rearrangement, forgetting sometimes that the relative stasis, not to say intransigence, of the printed book that we are familiar with is an anomaly in the history of the written word, and that user-made anthologies have been the norm.

Some complaints against print had more force. The earliest known call for press censorship was from a classical scholar, Niccolo Perotti, upset about Andrea de Bussi's shoddy classical editions being printed in Rome. Perotti wrote to the pope in 1471 asking him to establish prepublication censorship to ensure that texts were carefully edited. (The appeal was unavailing, and de Bussi became Vatican Librarian.) Similarly, the correspondence files of the Basel publishing house of Amerbach are full of letters to Johannes Amerbach contrasting his careful work with most printers' shoddy output.

The wisest people foresaw as well that the superabundance of books would lead to the promulgation of error. The famous chancellor of the University of Paris, Jean Gerson, in 1439 had complained that overabundance contained the seeds of error, and voices were heard as early as 1485 in Nuremberg lamenting not only the wide distribution of error but the uniformity and consistency of error in print. All copies of the printed book were alike and therefore it was impossible to compare and correct copies one with another. An error inserted in one was in all and there was no control as there was in collating individually prepared manuscripts.

Worse, an abundance of words would lead to confusion—of this, the *fourteenth* century was certain already. Nicholas of Lyre, in the second prologue to his literal commentary on the Bible, was mistrustful of even the limited hypertexting of the glossed manuscript page: "They have chopped up the text into so many small parts, and brought forth so many concordant passages to suit their own purpose that to some degree they confuse both the mind and memory of the reader and distract it from understanding the literal meaning of the text."

Printed books were marked by defects of presentation for a long time, often eerily resembling those of our own day: the scholar who has struggled to get the right Greek font for a computer's printer will sense a kindred spirit in the printer whose incunabular edition of Servius left blank spaces in which Greek words and phrases could be written in by hand. At what date contemporaries became sensitive to the loss of historical value in their treatment of manuscripts is an open question. Not until the mid-sixteenth century were large numbers of medieval manuscripts scrapped once they had been supplanted by print. Even then, to be fair, it was not their technology but their contents that rendered them liable to destruction: law texts and the Latin Aristotle had other reasons for obsolescence than the form in which they were presented. At the time of the Reformation, service books of the old liturgy similarly faced rapid destruction.

The most famous early critic of print is the immensely (if not always prudently) learned abbot Johannes Trithemius, of the Benedictine house of Sponheim near Frankfurt (later abbot of Würzburg). He wrote a book, printed in 1492, *de laude scriptorum* (In Praise of Scribes) that eulogized the old technology; years later in a chronicle of the monastery of Hirsau (1515), he praised printing, "that wondrous and previously unheard-of art of printing books." What do we make of this marvel of erudition in an age of transition who cannot make up his mind? Shall he be our patron saint of indecision?

Some of his criticisms make him out a typical foot-dragger. Anything prepared on paper won't last very long, he alleges. (There were already manuscripts on paper, but he is correct that the lighter, cheaper material was more commonly used by printers.) Writing on skins can last for a thousand years; but how long will print, a thing of paper, last? The shelf life of paper he estimated at two hundred years. The higher quality of the manuscript as an artifact and the value added by the scribes and illuminators also contrast with print. Scribes are more careful than the slapdash artisans of print, and so spelling and the other features of books are much more carefully looked after in manuscripts.

These criticisms of print that I have so far cataloged, both from Trithemius and others, are all true and valid. Every negative claim made about print is correct, and every negative prophecy came true. Take the argument about the likeness of copies making collation and correction impossible: a perfectly valid point. So why, then, did it not derail print in its glorious career? Let me suggest two reasons.

First, the point, though valid, is not decisive—that is, as much as we idealize correctly made books, we do accept a light admixture of error. The value of a book is not seriously degraded by its errors. Indeed, if that were the case, then the written manuscript book would also have been insufferable, but users were well acquainted by the fifteenth century with the praxis of deciphering imperfect books.

Second, and more important, the system of communication introduced by print was so wide-ranging, so fast, so powerful, and ultimately such a source of wealth that the defects of the system could be remedied as far as they needed to be. Proofreading was labor-intensive and wasteful in a manuscript scriptorium, but quite cost-effective in a print shop; and if the print shop is preparing stock prospectuses where large sums of money are at stake, proofreading of an obsessive-compulsive nature is both cost-effective and reasonable in view of the possible losses from error. If, moreover, the collection and comparison of errors is important for scholarship, then nineteenth-century scholars could develop the "critical edition," gathering and cherishing the variant readings of manuscripts, multiplying and then freezing them in print, thereby perpetuating whatever informational value they had to offer.

In the end, the defects of print and the criticisms they drew didn't matter. I want to suggest, however, that Trithemius in particular, was no mere Luddite violently opposing progress. The pieces of his criticisms of print that I have extracted here are only a small part of his whole treatise. His true topic is the undermining of the ethos of the monastery and its scriptorium. If in practice he approved of print and used it, he could still not find a way to bring print into his picture of the monastic life. The fact that he focused on preserving, not adapting, the monastic community was his failure.

Writing is the spirit's manual labor par excellence, and that way of life was threatened by printing. "In no other business of the active life does the monk come closer to perfection than when *caritas* drives him to keep watch in the night copying the divine scriptures . . . The devout monk enjoys four particular benefits from writing: the time that is precious is profitably spent; his understanding is enlightened as he writes; his heart within is kindled to devotion; and after this life he is rewarded with a unique prize." Then follows the story of a dead Benedictine who was such a passionate copyist that after they buried him, "many years afterwards," it was found that his three writing fingers were miraculously preserved while the rest of his body had rotted away.

The technology of writing had worked its way deeply into the social and economic structure of the monastic community. The monks who didn't know how to write were put to work binding, rubricating, making pens, and the like. To let writing go was to let something go that was perhaps not essential to the monastic ideal, but that had become integral to its practice.

And of course we now know Trithemius's fears were justified. Benedictine monasteries show a growth curve from 500 to 1000 C.E., ticking sharply upward toward the millennium, followed by lusty growth and sustained prosperity from 1000 to 1500, but the last 500 years—the age of print—have shown more mixed results. The social and cultural domination of much of European society by monasticism and its allied institutions faded rapidly in the sixteenth century. I make no determinist suggestion here, but only observe that the ability of the institution to survive depended on its ability to adapt itself to the new technological environment. (Universities did much better than monasteries, until now at least, though in the fifteenth century they shared many common traditions.) But Trithemius could not bring himself to theorize that adaptation.

He was not alone. A few places maintained scriptoria and print shops in the same house, such as an Augsburg monastery where the two coexisted from around 1471 to 1476, but most such arrangements lasted only for a short time. So while Heinrich Eggestein spent some time at the house of the Strasburg Carthusians printing the life history of their prior, Ludolf of Saxony, and teaching some of the monks the new technique, in practice, print was a business that flourished in less salubrious parts of town, among grubby businessmen unafraid of dirty hands. It was there that a new information order was created, and the social order found itself wrenched, sometimes agonizingly slowly, sometimes shockingly quickly, to align itself with what technology had created.

The most visible anti-Trithemius of our time was undoubtedly Marshall McLuhan who argued in the 1960s that media of communication strongly determine social history. There are other prophets still among us (think of Ted Nelson and his vision of a hypertext Xanadu), but McLuhan seized the high ground of public visibility at an opportune moment. Every lesson that Trithemius failed to grasp, McLuhan had learned at an early age. His lavish impatience with any criticism of new media of communication shocked his readers. His eager willingness to imagine large-scale social transformation far ahead of the curve distinguished his contribution to our common vision. He was both of his time and ahead of his time. If we do not yet live in a global village, we nonetheless live in a world bound closer by satellites and CNN than it was thirty years ago.

McLuhan's prophetic role represented his greatest success and his lasting failure. Though it may seem self-evident that new media of communication bring a powerful set of forces into play, McLuhan did not succeed in seizing the high ground of intellectual discourse, did not succeed in creating a line of successors, disciples, and pedantic periphrasts to follow him, and did not, finally, achieve the respectability that would have indicated that his ideas had been rendered harmless. In explaining this anomaly—that the prophet who most explicitly and, for the most part, most successfully addressed the conditions of knowing and communicating in our time is still so largely without honor —I will focus on two points.

First, prophecy is a mug's game—it is a style of truth-telling that assures it will not be believed. Aeneas could have told McLuhan that.

Although he had seen all the Roman future before him in his visit to the underworld, it did him little good. The prophets of Israel were famously ineffectual in shaping the behavior of their people. (Prophets say many untrue things as well, McLuhan his share of them, and that is no help of course, but everyone speaks untruths, yet we don't make that fact an absolute disqualification for credibility.) Prophecy is very gratifying to prophets themselves, especially if they have the fortune to live to the stage of seeing their prophecies come true (though often the prophet's personal status and even health may be made to decline at the point of verification: Cassandra found that out quite clearly). It is far less clear what social function prophets play. It is hard to know if they shape behavior usefully. The true usefulness of prophets may be for us to digest their theorized future after the fact, and it is early days yet to put McLuhan to this use.

Second, the intellectual domain that McLuhan inhabited is one that is unusually difficult to master satisfactorily. Scholars who address the kinds of questions McLuhan did—the impact of technologies on thought and behavoir—find it hard to conduct their discussion of the history of the conditions of intellection in a way that satisfies the prevailing criteria of intellectual discourse. The categories by which we do our intellectual business are so deeply ingrained in us that to turn our minds to relativize those categories, historicize them, and leave them intact but relativized, is, understandably, unsettling and disturbing. Consequently, to work our way into other mentalities is a fantasy at the outer limits of possibility. It is no more feasible than imagining that I am someone I am not. In the end, a McLuhanite reconstruction of technology-influenced mentalities of other times is and will remain a fiction by the terms of the system of discourse in which it is practiced.

The intellectual challenge of McLuhanism in all its forms is that it insists on asking us to transform ideas in ways that our own judgment tells us are illicit—to engage in a kind of magic, and therefore impossible, thinking about the past or even about the immediate future. We may agree, with McLuhan, that the newly literate Greeks were very different from the scholars of the era of the dawn of print, but we do not have the tools to bring those two systems of discourse in line with each other. We have only a system of discourse of our own, time-bound and technologically conditioned.

And so we fall away from McLuhan's visions unpersuaded, and rightly so, even as we accord him prophetic status and prophetic dishonor. If we are circumspect, we see that the underlying problem is in the history of mentalities. Whether self-consciously postmodern reconstruction of mentalities will prove any more successful is perhaps to be doubted. If it succeeds, it will have been because it begins so self-consciously, so self-doubtingly—the very feature that makes such forms of investigation and discourse so repugnant to the right-thinking cultural community beyond.

McLuhan's work is of great value, but does not have the value it seems to have. It is instructive, stimulating, and maddening—and perhaps most effective when most maddening. But its prophecies do not lend themselves to practical applications. Judged as myths, they are high-quality myths; judged as history or sociology, they fail.

So if we find ourselves in a whirlwind of conflicting ideas and new technologies, what then is a better way to proceed? Clinging cautiously to older social institutions is bad for those institutions themselves; bellowing prophecies into the whirlwind persuades few and leads to no concrete advances. Both roles have their important functions and will find practitioners, but we may be forgiven for pressing on to seek out a *via media*.

For my last exemplar, let me return to Cassiodorus, who suffered the indignity of serving as my dissertation topic and lent his name to my first book. I came to him in part because of his reputation for having snatched declining classical civilization from the barbarians, locked it up in the cloister, and taught the monks how to copy the classics—a Romantic image that I was later at some pains to demolish. I did not realize at the time that I had stumbled upon someone with eerie apposthis issues I have been discussing.

Cassiodorus belonged to that century or so of Latin Christian writers who were inventing Latin textual Christianity. He knew personally figures like Dionysius Exiguus, Eugippius of Naples, and of course Boethius, author of the medieval best-seller *The Consolation of Philosophy.* Toward the end of a long career as statesman, Cassiodorus had in mind a very conservative educational program of innovation: the establishment of a Christian school of higher studies at Rome, with the support of Pope Agapetus (535–6). He wanted to found such a school, not because he thought it time to unseat the classics, but because he thought the classics could take care of themselves, and that Christian textual study lacked funding. But war broke out, and Cassiodorus found himself an honored guest, that is to say a political refugee, in Constantinople, writing a commentary on the book of Psalms. When the war ended, he returned to Italy, not to any of the important cities where he had spent his career, but to his remote Calabrian estate of Vivarium, where he had founded a monastery on the family property. There, around 554 C.E., he picked up where he had left off twenty years earlier, with the same intellectual project that he had thought to pursue in establishing a Christian university at Rome. His *Institutes* are the intellectual schematic diagram of that project and a precious piece of evidence for early medieval Christian Latin culture.

In many ways, the project at Vivarium was a misfire. If it has been made out to be a turning point, it is because our narratives of the past insist on having turning points. There is no sign that Cassiodorus did a very effective job at inculcating Christian textual culture into his monks. He left them Pelagius' own presumptively heretical commentary on the epistles of Paul to be expurgated after his example (he did Romans himself, leaving the rest for them), and they made a cheerful hash of it, with the engaging result that this detailed commentary on Paul (still full to brim with Pelagian assumptions and interpretations, but sanitized of the most objectionable slogans) went forth into the middle ages as though it had a guarantee of orthodoxy about it. Because it boasted stray quotations here and there from Augustine, Augustine—through Cassiodorus—was made the unwilling and unconscious guarantor of the survival of Pelagius' ideas in this particular and pervasive form.

The last we see of Cassiodorus, he was still trying to train his monks as scribes by compiling a treatise on spelling, but the enterprise seems to have gone for naught. A few years after his death, the monks were squabbling with the local bishop, and their community subsided into oppressed obscurity. Enough copies of Cassiodorus' works survived to circulate in the middle ages, with varying effect. Not until at least one hundred and fifty years after his death did serious monastic scribal culture take root, and then it was in Britain; its spread thereafter was slow and uneven. In many important respects Cassiodorus was a failure.

But I have come also to see that this deflated savior of western civilization I learned to mistrust when I was young had nevertheless had the right idea. He did not despise the new; he used it wholeheartedly. He did not reject old social institutions, but rather found new ways to adapt them. He did not tarry to prophesy a new age of learning and wisdom.

Most of all, he *did* things. The larger scheme within which he did them was not widely imitated, nor was it imitable. Even to say that is to reveal what is so often wrong about our expectations of ourselves and our cultural heroes: we dream of strong leaders, knights on white horses, people who change history. Those are the fools and the demons of our past. The most effective change is wielded by those who do not expect to create or manipulate a closed system, but instead recognize that effective change takes place in open systems, where the accumulation of collaborative actions generates unexpected harmony.

The monastic Latin middle ages were predicted by no one, chosen by no one, built by no visionary hand. At a distance we can all argue how we could have built a better middle age. But that neglects the true merit of an age that achieved out of unpromising materials far more than it had reason to expect, and did so because it had stumbled upon forms of enhancing and institutionalizing autonomy and local responsibility—and if it is not obvious that I think here of the large social movements conventionally labeled feudalism and monasticism, then in just that failure of obviousness is our failure to imagine successfully how complex societies really are, how slowly they change, and how impressive coherent change of any kind really is.

So where does that leave us? By excluding the pragmatics of the old (Trithemius) and the theoretics of the new (McLuhan), I rule out two forms of behavior that academics in particular are fond of. What today's partisans of the book need to master is the pragmatics of the new. I suggest that Trithemius makes a good patron saint for our conservatives, and McLuhan an equally good patron saint for our theoreticians. In Cassiodorus, I find not a patron saint, but a colleague, a practitioner who innovated, failed, and innovated again. He did so on a scale and with a modesty of purpose that guaranteed he would eventually suffer the indignity of a debunking at the hands of a young whippersnapper; but an older practitioner of the new would at last recognize him as a colleague. Cassiodorus solved nothing: that is his virtue.

I mean by this construction no disrespect for theory, but perhaps a repositioning. When the lady Philosophia appears in Boethius' chamber, the Greek letters pi and theta on her garment and the ladder ascending from the former to the latter inscribe the precept that *Theoria* follows on *Praxis* and transcends it. A true pragmatics is not theoryless, but seeks the apotheosis of theory arising out of practice. The pragmatician is the person who hopes that at the end of the day the morning's theory will have been not vindicated but enhanced, even transformed, ready to reinvigorate practice and at the same time to be transformed again.

In this way, Cassiodorus used the modern codex book to display the novel kinds of texts his time had made. We cannot now reconstruct fairly just how novel, just how vexing, what he did might have seemed. The younger Cassiodorus seems as much a man of his time as the elaborately mannerist Ennodius, or the elaborately erudite Boethius. The older Cassiodorus went beyond their traditionalism; at any rate he abandoned their world of civil and ecclesiastical careers for a different kind of textual life. Would Boethius have gone to live in a monastery on the Ionian Sea? Ennodius, a social climber on the fringes of Boethius' circles, wrote his own educational prescription for young men of his time utterly classical, that is to say conventional, in all respects: to read that side by side with Cassiodorus is to see the difference between old and new in a single generation.

It also seems to me no coincidence that Cassiodorus is a name more readily recognized by graduates of library schools than by Ph.D.s. For the task he undertook, of imposing the most transparent possible intellectual organization on the body of texts before him, is quintessentially that of the librarian. In ages when knowledge was scarce, those who created it were the heroes of the tribe, and librarians their acolytes. But in an age of information overload, production and even dissemination of knowledge are child's play. Publishers hope that I will still be willing to pay for special pieces of information in the future, but I wonder if they are not too optimistic, not too much like Trithemius hanging on desperately to an obsolete social structure. The thing that I *will* be willing to pay for as the oceans of data lap at my door is help in finding and filtering that flood to suit my needs.

Among participants in the production, dissemination, and consumption of knowledge today, librarians have already made that kind of skill their specialty. They have, moreover, already led the way into the new information environment, for academics at least. They are caught between rising demand for information from their customers (faculty and students) and rising supply of that information and prices for it from their suppliers, and so have already been making pragmatic decisions about the importance of ownership versus access, print versus electronic, and so on. Can we imagine a time in our universities when librarians are the well-paid principals and teachers their mere acolytes? I do not think we can or should rule out that possibility.

Pragmatism is a difficult discipline. To reconstruct ourselves to fit the world in which we find ourselves is often a distasteful chore. Perhaps one way to make a fresh self-awareness easier is to make ourselves deliberately more conscious of the unnaturalness of this whole affair our culture has had with books. We long ago ceased to see the oddity of textuality and its institutions: publishers who produce books, libraries who treasure and make them available, scholars who pass on the mystic arts of interpretation to students. Is it not strange that we take the spoken word, the most insubstantial of human creations, and try to freeze it forever? Or try to give the frozen words of those who are dead and gone, or at least far absent, control over our own experience of the lived here and now?

Cultural continuity resides in memory, which is to say, in the keeping in mind of that which does not exist any longer. That is an extraordinary way to be human, and we may very well find someday a species in a remote corner of the galaxy that manages something like humanness without that elaborate construction of continuity. For now, perhaps it suffices to realize that everything we do in this line has something of a Rube Goldberg construction about it. Every now and then the complex constructions will be rebuilt, may perhaps even partially collapse of their own intricacy, and we will see that the genuine spirit of our culture is not in applying small pieces of cellotape to hold together the structure we received, but in pitching in joyously to its ongoing reconstruction.

In that vein, I would suggest that for all the passion and affection I bring to books, I have very little business caring for the future of the book. Books are only secondary bearers of culture. Western civilization (or whatever other allegorical creature we cook up to embody our self-esteem) is not something to be cherished. Western civilization is us and making it, as well as remaking it, is our job. The thought that we come here in a generation surrounded by opportunities to botch the job might be frightening—or it might better be exhilarating.

Hyperlink

WHO OWNS THAT IDEA?

The O'Donnell family does not have an unblemished record in matters of copyright law. In the sixth century, one of our number by the name of Columba, wanted to own a copy of the Psalms in Latin and so hand copied one from a manuscript belonging to another man. The owner of the original copy objected and took the case to the local king, who decreed that "as the calf belongs to the cow, so the copy belongs to the book" and awarded custody to the owner of the original. Columba resisted the decision, and a skirmish broke out—something between a full-scale war and an Irish pub brawl. In the end, Columba held on to the book. However, Columba's confessor disapproved of the violence Columba had used to retain possession of the holy book and so compelled him in penance to leave Ireland and never set eyes on the country again.

That's a fine old story and quite well attested, but it is anomalous only because of the time and place from which it comes. The clumsiness of the king's ruling, with which I think most of us disagree, reveals the absence of any clear or consistent notion of intellectual property. For while that phrase is an oxymoron, it is a powerful conception that has determined much of the way we produce, distribute, and consume ideas of all kinds today.

The necessary condition for emergence of that idea was the invention of printing and the possibility of producing multiple copies of a single original for sale quickly and cheaply. The relation between the writer of the words and the owner of the press was bound to become problematic.

(That it did not become so at first is because the invention of printing did not produce much new literature at first. The vast majority of the earliest printed books were written decades, centuries, or millennia earlier.)

For a very long time the old pattern of support for writers continued: patronage, that is to say, subsidy of the writer by some wealthy person or institution. Copyright as we know it (including the word itself) is an invention of the eighteenth century and accords, not coincidentally, with the moment when Samuel Johnson famously became one of the first English writers to support himself on the proceeds of his labors in the absence of patronage. By the time of the American Constitution, copyright was important enough that Congress was given the power to manage it "to promote the progress of science and the useful arts"—a progress that depends on authors' being encouraged to write and publish, and on readers' being encouraged to read and react to what they read.

The idea that words could be property and that the author could control where, when, and how they were reduced to print gave authors a powerful place in the economics of the printed book. At the same time, the contrary principle embedded in the earliest copyright laws and in others down to the present time, that owners of books really do own them and can dispose of them as they will, turned out to be equally empowering. If economics may be reduced to supply and demand, the genius of the copyright system that emerged in the eighteenth century is that it did justice to both sides of the equation. It created a way for authors to be compensated for their words, and it created a way for readers to use those words freely and responsibly. It made possible the growth of large public and institutional libraries, where books are shared among a community that could never afford to buy all the copies they need of individual books themselves.

But the idea of copyright still depended heavily on technology. Copyright is easy to enforce when copyright can be violated only by someone with a large capital investment in a printing press. Under this constraint, it was possible to violate copyright and get away with it, but risky. If the result of such a violation were confiscation of a substantial part of the violators' capital assets, then people would think twice about doing it.

Whether copyright will survive in a largely electronic environment is a subject of heated debate. The technological underpinnings of the idea have been gradually etiolated to the point of irrelevance. The photocopy machine started the onslaught, letting private individuals with access to a moderately expensive special-purpose machine circumvent the requirement of purchasing the printed copy. In the thirty years or so of the history of xerography, we have seen the threat of unrestricted copying checked by copyright violation rulings against copy centers. Surely a greater percentage of copyrighted information is illicitly copied today than was the case a generation ago, but the information economy has adjusted to that practice and continues to thrive.

But the electronic media seem at the outset to offer an even greater threat to copyright protection. The disadvantage of the photocopy was that it was cumbersome to make and cumbersome to use. No one prefers, if they have a choice, a bundle of toner-stained paper to a finely bound and printed book. But if information exists as a pattern of 0s and 1s on a hard disk somewhere, to produce a copy need not be more than a matter of nanoseconds, and the resulting copy may be functionally identical to the original in every possible respect. Worse, the technology needed to make such copies can be relatively cheap, lightweight, easily concealed, and difficult to trace. Has the barn door been torn from its hinges?

Perhaps, and perhaps not. Human ingenuity is great, and where there is a desire to protect information, there is every reason to think that considerable success can be achieved. If information depends on something about the particular platform where it resides (the software it uses, or a structure of files in a directory) and is thus cumbersome to remove, and if pricing is reasonable, then few if any readers or viewers would have the necessary incentive to remove it. Just as with photocopied information, the challenge is not to prevent every theft, but to discourage theft sufficiently to leave an economic base for the primary providers of information.

It is not therefore useful to ask whether copyright will survive. It can, as long as providers of information want it to survive. Indeed at this moment, the real risk is that providers of information will take advantage of their leverage with government to place greater restrictions on the flow of their information in cyberspace than has been the case regarding print information. Legally binding license agreements that readers sign to get access to a database already require them to behave in ways far more restricted than copyright law would ever have demanded. Further, there is a major assault on the legally defined concept of "fair use"—the kind of copying and republication of information for scholarly and scientific and educational purposes which goes somewhat beyond what the ordinary reader may be allowed.

These forces threaten our public and research libraries as nothing has in decades. Today, libraries can buy an encyclopedia and put it on the shelves for anyone to use. Tomorrow, they may have to pay a large fee to get the encyclopedia, then be charged an additional fee for every use of the resource, be forbidden to let anyone not a member of a particular university community have any access to it at all, and be required to give it back if they stop paying an annual fee.

These are worrying developments, no question. But they do leave room for optimism. Excessive restriction breeds demand for unrestricted access, and those publishers who hold on most tightly to their product may find it rendered obsolete by more freely accessible competition. In addition, the emerging culture of the Internet, with technological possibilities very different from those of the print world, may well find new ways to envision the economics of information.

The latter possibility is genuinely exciting. The culture of the Internet is marked by a circle-of-gifts mentality, according to which people produce materials and contribute them to a common stock on which they draw themselves. The technical threshold to "publication" is far lower than it ever has been before. A computer and a modem can enable you to distribute thousands of copies of your words to the world in a matter of minutes. Someone who has put time and effort into creating an information resource (a list of all UFO sightings in central Idaho in the last fifty years; or a minute study of the theological bent of a set of fragments from the Dead Sea Scrolls—high purpose and low can mix easily in this environment) needs no intermediary to distribute that resource in fully functional copies to the world. As online communities of interest emerge, they become sharing grounds for this kind of information.

There are already serious proposals afoot to use this aspect of networked information to distribute scholarly and scientific information of high quality. The market by which this scholarly and scientific information has until now been distributed in print is already a somewhat factitious one, since most of the costs of producing the information (that is, the costs of doing the research and the writing) are heavily subsidized by institutional employers. The percentage of the costs involved in distribution to the world is tiny by comparison. What if, the argument goes, only a small shift in allocation of costs can make such discourse freely available to a wide public faster and cheaper than ever before? If that vision becomes reality, then copyright would become obsolete for science and scholarship and a whole new conception of what intellectual property and how to regulate it will be needed.

A reasonable user of information today observes and obeys copyright and license agreements punctiliously, but at the same time looks for and thinks about opportunities to create alternate information economies for the common good. Networked electronic technology has immense potential for democratizing access to information—that same computer and modem can bring the world to your doorstep, even if your doorstep is thousands of miles from the nearest university library.

The book that Columba copied went with him to Scotland and stayed in the O'Donnell family for over a thousand years. Columba was venerated as a saint and the book was considered a relic: an ornate jewel-encrusted case was made for it, and the whole was regarded as a good-luck charm. Called the Cathach (Battler), it was carried into battle by the O'Donnell clan for almost a thousand years, until it was given into safekeeping in the sixteenth century.

The ultimate irony of the story is that in the end, the Cathach turned into a piece of property which had lost its intellectual value. In the nineteenth century, the case containing the book was returned to Ireland by a member of the family who discovered with some surprise that there was a book inside at all. The good-luck power of the artifact and the beauty of the case in which it was enshrined had obliterated from memory the very fact of the book's existence.

And that is the risk in intellectual property laws: based on an oxymoron, they can so thoroughly transform intellect into property that the original purpose is quite lost. When American trade negotiators took the Chinese government to task for violation of copyright in 1994, it was not over theft of materials that could reasonably be thought to promote the "progress of science and the useful arts," but over punk-rock CDs and *Lion King* videos, the devices commerce has contrived to turn intellectual property into vast profits. The sagest course for those who care about the quality of their discourse and who cherish truly intellectual property is to look hard at the new media for ways to keep free and open economy in ideas, while letting the idea-less thrash each other with lawsuits and threats of trade wars over cartoons and noise.

Chapter Five

THE ANCIENTS AND THE MODERNS: THE CLASSICS AND WESTERN CIVILIZATIONS

A. E. Housman, known for his slender volumes of sensitive verse, was a late-blooming classics professor. His brilliant undergraduate career at Oxford went off the rails at the last moment, and after failing his exams, he spent a decade in the wilderness working as a clerk in the British patent office, continuing his classical studies at night, and publishing scholarly articles. His work was of such high quality that he was hired as Professor of Latin at University College, London, at age 33. After almost two decades there, he moved on to Cambridge, where he divided his life between the austere pleasures of scholarship and the pleasures, for him scarcely less austere, of the table.

Housman was ferocious in his attention to detail and in his disdain for those who failed to meet his standards of learning and accuracy. He is perhaps the most famous representative of a long tradition of a particular form of scholars' bad behavior for which we even have a Latin name: *odium philologicum,* or philological hostility. Housman's own work almost redeems his bad temper by its rigorous fulfillment of promise. He specialized in editing, or discussing the editing of, Latin poets. His editions of Manilius, Juvenal, and Lucan remain important scholarly tools to this day. And scholarly tools they are; Housman did not write for the general public, and his editions boldly claim on their title pages to be only for the use of other editors. Of the three editions he achieved, only the least impressive, the Juvenal, handles a poet who has much resonance outside specialist circles in the twentieth century.

In any potted description of Housman's work, this is the point at which tradition demands that one regret the squandering of such rare poetic talent on obsessive-compulsive preoccupation with poets less talented than himself. But such preoccupation with the individual fails to do justice to the authenticity with which Housman lived up to a type implicit in his culture. To be Professor of Latin or Greek at Oxford or Cambridge had, in the nineteenth and early twentieth century, a prestige comparable to that of the cancer-fighting geneticist in our own time. The combination of madness and wizardry that Housman represented could have come about only as a result of the institutions and obsessions that shaped him. (I regard Housman as the English Nietzsche—similar in that both were paragons of classical scholarship, and both went mad because they took themselves and their classics with such deadly seriousness. Zarathustra is the Shropshire lad's extraverted cousin.)

The classics dominated the literary and educational landscape of Housman's day. The recovery of contact with the Greek world in the late eighteenth and early nineteenth centuries (as the Ottoman grasp on the Greek homeland began to weaken) contributed to the creation of a century of intense activity in classical scholarship. To the generations after Goethe and Winckelmann the return to the Greek and Latin classics looked and felt like a liberating flight from bourgeois Christian society into a world of saner, cleaner, brighter values. But what had been enthusiasm in one generation became scientific discipline and hard, positivist recovery of the truth of the ancient past for the next. The culmination of this spirit could be found in the Berlin scholar Ulrich von Wilamowitz-Moellendorf, who combined dazzling scholarly reach, minute attention to detail, and a flair for interpretation that made his public lectures the cynosure of the ladies of Berlin.

But the éclat of the classics ran beyond the specialist even in Britain. One of E. M. Forster's teachers at Cambridge, Goldsworthy Lowes Dickinson, was so imbued with the liberating spirit of classicism that he wrote his own dialogues in a Platonic vein. The Professor of Greek at Oxford when Housman had a chair in Latin at Cambridge was Gilbert Murray, who won a wide public for his reassuringly Edwardian translations of Greek tragedy into English for acclaimed production on the British stage. They read today like period pieces, far more instructive about Murray and his times than about anything Greek.

Upper class boys in those days could expect to spend years of childhood and adolescence learning Greek and Latin. If we could today examine them in person, their knowledge would amaze us by its accuracy and its unsophistication. They surely knew their ablatives and periphrastics better than most graduate students today know them, but they read their texts with a plain literal-mindedness and obtuseness that is enthralling to behold. Though the boys often showed little enough interest in their studies, the crushing repetition (backed up by the threat of corporal punishment) meant a certain inculcation of substance was bound to occur. In those bright enough to go on to the universities, a fair amount stayed. (One of Housman's editions is dedicated to a sentimental friend of his undergraduate days, no classicist and indeed a scientist by training, but the dedication is in many elegant lines of elegiac Latin verse, which the recipient was clearly expected to be able to understand.)

But the classics penetrated beyond the elite strata of that society. Consider Rudyard Kipling. Though Kipling has been rather rehabilitated of late, and the bookstore shelves have a long run of titles of his books, the one that is perhaps hardest to find is the one I have found the most instructive: *Stalky & Co.* This volume appears to be merely a collection of loosely connected stories about British schoolboys of the late nineteenth century, at a not very distinguished public school, swotting up their Greek and Latin and falling into mischievous adventures. First published in hard covers about 1899, the book reappeared slightly expanded with a few more stories and a few verses after World War I.

The most telling of these tales is one that is only in the expanded version of the book, a story called "Regulus," which begins with the class hard at work translating a difficult piece of Horace's poetry, the story of a singularly heroic and self-sacrificing Roman. It then continues with a small drama of school life in which an unworthy victim suffers unjust punishment—and gradually it dawns on the reader that life is recreating art on a tiny scale. Many things were done with the Greek and Latin classics in Victorian England, but their role as improving models and sage examples of what to avoid was paramount. The particular intersection of imperialism, manly virtue, and veneration for the ancients is a distinctive late Victorian creation.

This education by exemplar reaches its high point in the last chapter of *Stalky & Co.,* when we leap forward from the school a few years to a tale of life on the Indian frontier of empire. At this point it becomes clear that even in 1899 these were tales of a burnished past, for Stalky's adult adventures are set in the 1880s. (For this book had its life first in the dark days of the Boer War and was republished after World War I, when the ambiguities of empire had been brought home to the British, but when it was still possible to tell these stories of unabashed Victorian imperialism.) Gradually as the narrative continues through its few pages, the attentive reader begins to notice that every detail of Stalky's derring-do in the British Army has been anticipated by some reconnaissance, raid, subterfuge, or bold defiance in the schooldays of a preceding chapter. In a way the whole book is a gloss at the end of the British imperial century on the Duke of Wellington's famous *mot* from near its beginning about where the battle of Waterloo was won. Through that period, the ideology of British educational self-understanding was classical.

This is not to say that classical education was unilaterally imperialist. It could hardly have been so pervasive if it were. Rather, its pervasiveness exercised a more subtle effect on British culture, offering a limited variety of highly constructed models for thought and behavior. In reverse, it should be no surprise that the works accounted the classics were subtly modeled on the Britain they illuminated. Particular periods and authors had their vogues in which prevailing lines of interpretation made clear the suitability of the ancient authors for modern use. The elision of the homoerotic qualities of some of Plato's works (including the *Phaedrus*) was incomplete and, to some eyes at least, transparent, but sufficient to satisfy the decorum of a society determined to suppress or at least deny the homoeroticism in its own midst.

Nor were the classics the only source of authority: far from it. The nineteenth century may indeed have seen the apex of the notion that salvation and wisdom lay within the covers of a book, and the Christian scriptures retained a powerful hold on the imagination and usually the belief of even the most classical of scholars. What is remarkable is the nonchalant syncretism of ideologies and the transparent expectation that Greece, Rome, and the Christian Bible all supported and justified the Britain of that time, a belief none the less remarkable for being shared (with the necessary nationalist adjustment) in several other countries of that day.

If one may venture, as something of an unbeliever in this regard, to give voice to the creed that underlay that nineteenth-century classicism and which still, suitably ironized, lies behind much of our contemporary debates about educational content, it would go something like this: In the beginning the earth was formless, void, and prehistoric. Civilization began in Mesopotamia and Egypt, but their writing systems were obscure and therefore inferior, and their societies ridden with priests. The Greeks then acquired both alphabet and democracy, and wisdom and great art ensued. We are prepared to overlook their many priests. But the Greeks were too impractical to thrive and so subsided as a culture, giving way to the Romans, who brought single-minded practical skills to bear on the question of organizing government. They succeeded in establishing a peaceful, prosperous empire across the known world. The Roman empire lasted five hundred years and then fell, for reasons that ultimately can be traced back to some inner failure. What ensued were the dark ages. The dark ages have been somewhat redeemed in recent decades by expansionist takeovers from both directions, and healthy swaths of that territory are now either late antique or pre- or proto-modern, but the expectation is that they remain a period overtaken by religion (we are not prepared to overlook *their* priests) and a near lapse into prehistoric inarticulacy. Fortunately the light dawned, and the dark ages were followed by the Renaissance, then early modern Europe (a providential preparation for ourselves), the Enlightenment (better and better), romanticism, modernity (with hints of dissonance), and now perhaps (a bit nervously) postmodernity.

There are many things wrong with this narrative, though in one way or another it is the master narrative of all discussion of western cultures today, even for those who dispute it. Since there is no credible substitute, most critics wind up disputing the value of parts of the story or rewriting bits of it. Still the master narrative remains, disbelieved, but in place.

The most emphatic failing of this linear narrative is embedded in the narrative itself, namely, that the line of succession was broken and needed repair. The presence in modern European culture of the Greek cultural ancestors in particular is a willed choice, made in stages repeatedly between the fourteenth and the nineteenth centuries. A millennium of western history had passed without paying serious attention to the Greeks, indeed had gradually parted company culturally and ideologically with living Greeks to the point of mutual excommunications in the eleventh century and outright warfare at the time of the Crusades. Whatever the value of the Greek accomplishment, historians in the western middle ages only notionally believed in that line of ancestry and preferred to emphasize their biblical descent from Adam through Israel and Christianity, with Rome as a convenient vehicle for the expansion of Christendom.

The decision to bring back Greece was a conscious one, made by the most "advanced" elites of the times. Think of Byron's passion, or the glamour of the Greek "Orient" in *The Count of Monte-Cristo.* Furthermore, the construction of this narrative itself came only in stages. It is a lexical fact, for example, that the word "Renaissance" appears in English and French only in the nineteenth century. To be sure there was language of rebirth and renewal in the air in the fourteenth and fifteenth centuries. (Indeed those were periods of far more than rebirth. A serious assessment of their excellences would give them credit for far more innovation than mere restoration.) But it was that text-crazed nineteenth century again that needed the Renaissance to be part of its own cultural legacy, its own construction of what its past and its legacy had been.

Readers of John LeCarré will recall how his story *Smiley's People* begins with the Soviet spymaster Karla making a legend for a girl—cooking up a story about who a mysterious girl was and who her parents had been—to justify her existence and protect her from hostile forces. In one sense, the potted history of western civilization I gave a few paragraphs ago is just that sort of legend.

But lest it seem entirely arbitrary, ask yourself this: who are your ancestors? If you are like me, you have instead of a census register, a few stories. I can carry my stories back to the late nineteenth century in the west of Ireland in one direction, and as readers of this book have found, I love to tell of the great deeds of the O'Donnell clan all the way back to the fifth century C.E. But when I do so, I am making up a story by selection, choosing one line of ancestry over a multitude of others. My Polish grandmother and my English grandfather enter nowhere into those stories, though fully half of what I may claim to be is their gift. But I tell the family story that I need to explain myself to myself and to others. We all do that as individuals, and we do that as a culture.

Thus when we quarrel among ourselves about things like an American history curriculum, we should acknowledge that very little science or scholarship is at stake. Rather we debate the construction of our national legend, which shifts and changes shape over time. George Washington and Abraham Lincoln have subsided noticeably in my lifetime (look in vain if you doubt this for looming portraits of Lincoln at the next Republican Party convention), and there are fewer heroic war stories told in our classrooms since Vietnam.

So whoever "we" are, our version of western civilization is selective in the extreme. A useful comparison here is with China, whose attachment to an ideographic writing system and to a small set of ancient classics had the practical political effect of shaping a common consciousness of China that held together far-flung peoples and cultures. These classics' story of "Chinese civilization" proved to be a self-fulfilling prophecy to a far greater extent than ours. But they had no great interruption of a thousand years in their telling of the tale about themselves (as we did during the "dark ages"), and they maintained an inclusive view of their selfhood for millennia.

By contrast, the distinguishing feature of the western legend is its diversity. Greeks, Romans, Jews, Germans, Celts: the would-be homogeneous past we create for ourselves is already culturally diverse in the extreme. It is also selective. Though we get our religion from the Jews, our alphabet from the Phoenicians, and our numerals proximately from the Arabs, our civilization is distinctly anti-Semitic or at least un-Semitic in its self-image. The classics were chosen to be those of Greece and Rome, and those two cultures remain yoked together in our curricula while others are kept out. (Romans and Germans have a similarly long history of co-dependency, but there are no academic departments of Latin and German that I know of.) It is also emphatically true that Greeks, Syrians, Jews, Arabs, and others created in the near east a large and vibrant culture of extraordinary diversity, but one that, like the Semitic cultures, has resisted satisfactory study until quite recently, owing to the variety of competences not ordinarily inculcated by our schools that it requires.

In a sense, we chose our ancestors, and our history saw to it that they behaved accordingly. Ours is a culture imbued with Greek and Roman ideas, images, and for that matter architecture, not because the Greeks and Romans are our ancestors, but because we have chosen them to imitate. There is a serious history that could be written of direct cultural filiations and continuities with antiquity, but this would not boast the poems of Horace or the philosophy of Plato among its high points. It would need to be a history of technologies and cultural practices. There, to be sure, both Greeks and Romans would have important roles to play in our history, but so would many other cultures. The story that would emerge would not resemble very closely the legend of western civilization.

Legends like this do not merely happen. That our western legend is a restoration should not lead us to think that the story is true and complete, but that our reception of it is what has mattered. Here again the comparison with China is instructive. The "Five Classics" at the heart of Chinese culture did not emerge naturally by common consent. They came hedged about with founding legends, not least the ones that transparently seek to attach the name of Confucius to the transmission of each. We must distinguish between the moment of creation and the equally important but usually obscure subsequent moment of selection and canonization.

The same is true, dramatically, of the Bible, a collection of books written at many dates and times, with a long history of gradual coalescence and authorization. Attaching the name of Moses to the Pentateuch was one authorizing gesture; giving the name of David to the collection of Psalms was another; while the book of Daniel grew over time. The collection of Hebrew scripture as a whole that was received in the Greek-speaking world had its founding legend in the story of the translation by the Seventy Translators (the Septuagint), who went into their separate cells and came out months (or years?) later with miraculously identical versions of the same texts. Whatever else it is, this is a story designed to explain and justify a particular collection of books in a particular version to its readers.

Similarly, we think we know the story of collection and authorization of the Greek and Latin classics, because we know the literary histories of those cultures, but those histories as commonly told do not go far enough. The texts we receive as Greek and Latin classics date to remarkably narrow periods. For Greece, the most familiar authors after Homer lived, for the most part, within a century of each other. For Rome, the period is a little longer, but the golden age from Cicero's fight with Catiline to the death of Tacitus the historian is still less than two hundred years and includes some dry spells.

To turn them from contemporaries into classics was a subtle process. The writing down of Homer had made him central to the Greeks, and Vergil's self-consciously imitative and derivative *Aeneid* was drawn into the classroom early. On the Greek side, the Hellenistic age (roughly from Alexander the Great through the Roman empire) saw a gradual refinement and clarification of the Greek canon of authors and texts that would abide in Constantinople through the middle ages.

More interesting was the case of the Latin classics. Latin literature, flourishing to around the turn of the second century C.E., largely died out as a living practice in the second and third centuries. This reflected the inanition of the literary classes in the Latin west, in contrast to the continuing liveliness of the (always more culturally ambitious and focused) Greek east in that period. The fourth century C.E., the period that saw the establishment of Christianity as the official religion of state, was also the age most responsible for exhuming and classicizing the Latin past. Authors who had not been read for centuries (as far as we can tell) were suddenly read and copied.

The artificiality of the exercise is best seen if we look at who was doing the reading and copying. Seen from the outside, the literary class of the fourth century was unchanged from that of four centuries earlier. It was the senatorial aristocracy of the city of Rome, every family with its own legend and its own collection of ancestral masks in the central hall of its house. But those legends were bogus to the core, and the vast majority of those families were parvenus, lately ennobled, and wholeheartedly buying into the dreams of Roman glory that their political and financial advancement encouraged. By a stroke of genius, the emperors of that time reinvigorated their regimes by creating this new class of old families, so to speak. These were the people who could see perfectly well that reading the old books was the sort of thing they should do, and they did it: pedantically, to be sure, short-sightedly, without a doubt, and without a tithe of the creative energy and spirit of those they read. Their choices of favorite authors determined for the most part what survived into the middle ages and thus the content of the Latin "canon" to this day. (Though there are, to be sure, authors who survived in spite of neglect in that day.)

The irony here is that the creation of the classical past, that is to say, the selection and the ennobling of it as model, was a postclassical achievment. In the case of Latin, it was the same age that both adopted the revolutionary, anticlassical Christianity and created as well the classical past. Much of that classical veneration of the Latin past survived the middle ages intact, and its classicism was itself one reason why, when commerce with the Greek east was possible again (mainly when the Greek world had fallen under Islamic control), the idea of classicizing the past and of privileging the Greeks alongside the Romans seemed so familiar. The final irony is that the "Renaissance" only did what the easily dismissed dark ages would gladly have done.

When we come to modern arguments about the nature of the classical past, it seems that we are seldom far from antiquity itself. That is the most seductive thing about the study of the classics, and the most deceptive. The staunchest defenders of the classics regularly argue that Greek and Latin literature and culture are so inextricably interwoven and mutually bound that it is impossible to study the one without the other and that taken together they amount to an achievement of a very high order. There is much truth to this argument, but it needs to be handled with care. That mutual referentiality is very much the result of conscious choices made by ancient authors. Hellenistic and Roman authors deliberately linked their own creations to those of classical Greece and thus deliberately built a body of literature that was smooth and whole and complete. It insists on being taken whole if we are to understand a part of it.

And so for pragmatic reasons, we must read Homer to understand Vergil, Plato and Demosthenes to understand Cicero, the Greek lyric poets to understand Horace, and so forth. Or at least to understand them in the way in which they were created to be read—and there we give away the ball game.

In the explosion of varieties of literary criticism and cultural study in the last thirty years, we have learned that there is more than one way to skin a cat. Faithful, devout reading of Vergil has its place, and helps us understand not only Vergil but also Homer, Dante, and Milton in the bargain. But quarrelsome, insurrectional reading of Vergil ought to have its place as well. We don't need to accept Vergil's view of himself and his poem—indeed we shouldn't accept it—as a precondition to all reading of it. Too often we read the classic poems in a landscape of implied literary historical narrative that cannot possibly withstand contemporary scrutiny. To break up that landscape and find multiple paths into and out of Vergil is to enrich our sense of what he was and what he still can be. Perhaps, given the emphasis in the *Aeneid* on Latin *pietas* (a virtue of fidelity to one's conventional social obligations), the best prescription for the future study of such texts is a little more disciplined impiety.

For one brief period in the eighteenth century, there was a chance for it all to have been different. In France the "Querelle des anciens et modernes" and in England the "Battle of the Books" erupted in the 1690s and burned a line through the polite studies of the day.

The controversy now impresses us by the eminent reasonableness of the side that seems to have lost. Its partisans held that the function of literary studies and the life of the mind in general was the improvement of the student and his society. On that fundamentally pragmatic ground, they held that the ancient authors should yield place to the moderns. For science progressed, and moderns knew more than ancients did: to continue to defer to Aristotle in philosophy was as foolish as to defer to Galen in medicine. The battle raged for a generation (Pope's *Dunciad* is its tailpiece), and in the end the moderns both lost and won. On the one hand, conventional educationists prevailed. The classics remained central to education, and the way was paved for nineteenth-century innovation and insecurity to settle into their study. On the other hand, what was changed was fundamental. Tacitly all sides agreed that the classical authors were no longer fit models in all areas. The "two cultures" that C. P. Snow found coexisting and rivaling each other in our time really owe their ancestry to this period, when the practical sciences forswore their links to antiquity for the most part, and the literary, philosophical, and polite sciences insisted on retaining them. The classics thrived again, but as objects of a study whose ability to improve modern life was at best unproven. The argument that reading Plato or Tacitus improves the mind or the soul is, after all, hard to prove.

One further contingent cultural event, however, conspired to help guarantee the classics a place in modern life. The eighteenth-century rebellion against Christianity that we call the Enlightenment was a deliberate attempt to make a place in the cultural landscape for a non-Christian view of life. The philosophes of that age who sought a veil of authority for themselves were ingenious enough to see one in the past, the classical past. What had been implicit before, they made explicit in their own work. The return to the classical past was taken not because the ancient past embodied perfect wisdom, but because it embodied wisdom that was not Christian.

The reading of the classics that the eighteenth century perfected and handed on was one in which serene, skeptical, mildly anticlerical men of letters passed ironic judgment on the foibles of a world given to folly and religion. That view of the classical past was in many respects a parody, but it was instantly persuasive because of what it offered the world in which it was proposed: an intellectually respectable alternative to Christianity. Melded with Christianity in the nineteenth century, it provided alternate ways of thinking about virtue and public affairs. In the end, the classics of the eighteenth and nineteenth century were the creation of the needs and desires of those periods. It can be argued (with only slight exaggeration) that classical antiquity was created in the eighteenth century and is one of the most impressive creations of that age.

The western civilization that created a classical past (and the dark ages) to explain itself can probably be identified with what some call the "long nineteenth century." From the French Revolution to World War I, the liberal society of northern Europe arose, sustained by its delusion of a long happy peaceful future for its dominance of the world stage, and collapsed in 1914. If we still have not come to grips fully with the collapse, it is because part of us still wants to live in the delusions of that age, and part of us wants to rid ourselves of the truths of that age as delusions. If we could see Marxism more clearly as the apotheosis of the nineteenth century's optimism, an optimism that our most conservative politicians continue to embody, we would come closer to seeing a way ahead. As it is, it is a usefully gloomy fact that the failure of the dream of western civilization has left us a culture with no ideas. I say usefully, because it is at least arguable that ideas kill more readily than anything else does, and that the assorted dreams of a world beyond ideology offer our best hope of a pragmatician's future.

Just who we exclude and include in our cultural legend-making has lasting political effects as well. Beyond the traditional bounds of Europe lies the great alternative within western civilization, the Byzantine-Orthodox-Russian tradition. It is scandalous that we keep Greek and Slavic Europe of the last millennium so far from our cultural concerns, and our obtuseness as a culture for the last fifty years in dealing with the Soviet and post-Soviet states of that world is owed in no small part to that cultural purblindness.

This willful ignorance is second in potentially calamitous effect only to our rejection of Islam, which we do not allow even to be "western." To be sure, Islamic traditions raise the question of the usefulness of the category of western, but it is little short of astonishing that a culture that has been a constant part of the European political scene (first in Spain, then the Balkans) for thirteen hundred years should be so resolutely rejected as alien and eastern and distant. The price is paid for that folly in Bosnia now virtually every day.

The United States of America became aware of itself as a culture in a strictly European environment. All our important relations for two hundred years were with European powers. World War II was a shock: an encounter with a genuinely Asian power. It was a shock that seems to have taught us nothing. We Americans still struggle to live in an Anglophone world, barely scraping up a familiarity with one or another European language. We have the preposterous chutzpah to think it a *disadvantage* if our society becomes more multilingual because our insularity is threatened by it.

And we live with a self-image decades if not centuries out of date, thinking of ourselves as young, free, uncorrupt. When our economy shows signs of running a little less roughshod over the rest of the world than has been our wont, we sink into recession and imagine prospects of gloom. The only narratives we can imagine are of triumph or of decline and fall. We do not see that we have aged and become mature, and that we are now to the world what Europe was a hundred years ago—triumphant and vulnerable to upstarts. So too the American west is full of people who think they are themselves still brave frontiersmen, when they are only their etiolated descendants, too unimaginative to reinvent themselves.

It is impossible to go to Hong Kong today without thinking that the upstarts have arrived. Europeans of a hundred years ago coming to the United States had the same experience of being astonished into inarticulacy by the energy, the ambition, and the scale of the economic whirlwind into which they stepped. To be sure, there are threats that could derail *any* prosperous future for any society, but suppose for a moment that the western Pacific rim continues to prosper and grow, and suppose again that the billion-and-more inhabitants of China are brought successfully into the world consumer marketplace. What then? What will the history of the world look like a hundred years from now?

Will sober and reasonable analysts, whether they write books, speak on screens, or project ideas telepathically, not perhaps tell this story? That for two thousand years the wisdom of Asian civilization was sure and reliable; that the intrusion of the bumptious westerners in the nineteenth century threatened for a time to unseat Asian traditions, culminating in the fascist boondoggle of World War II Japan and the Stalinist tragedy of China; but that in the end the truly superior cultures prevailed, absorbing the best of the western lessons but maintaining the virtues of the eastern cultures, and coming to dominate the world's economies with benevolent authority.

I do not tell that story to commend it but to ask, what if that becomes the most compelling legend about our time? Must we fight or surrender? Must we stick up for western civilization or be swallowed up by the Asian horde? The deadliness of the alternatives should be the best warning that the problem is badly posed. The underlying assumption of fundamental opposition between eastern and western cultures can very well become a self-fulfilling prophecy, if we are silly enough to believe it. To believe in a divided world is to create a divided world. If we remain western and they remain eastern, then it is reasonable to think that both sides are doomed.

Instead, perhaps east and west will find their way to a better common culture that transcends the provinciality of those ancient but lethal imaginations. We do not have to continue to behave as though we were from different worlds. If we will finally learn Columbus' lesson, that one reaches the east by going west, the opportunity will be there to make a more interesting and more convivial world.

To create this opportunity, we must first show more imaginzation than we currently do in our own legend-making. As long as we fight about which legend of western civilization we choose, we betray the past by using it as a projection of ourselves, and we betray the present by imputing our problems not to ourselves but to our past. The topic of western civilization is controversial today for two reasons. The first is purely ideological, tied up with debates about the present fought out in terms of the past to be selected. Do we live in the best of all possible worlds, where things get better all the time? Then western civilization is the heritage to cherish. Is the world we inherit fraught by insufferable tensions? Then western civilization is the villain. (The curiosity here is that all parties who have an agenda agree to use this quite artificial construct as the focus of their contest.)

The second reason is that we can't think of anything better: the continuing vitality of the topic is itself a sign of a failure of imagination. What is the function of recurring to ideas and debates of two thousand years ago? Is it not reasonable to suggest that we are gleaning well-reaped fields? Is that *all* we can think of to do? Look at the arguments we fight through. In recent years, for example, classical and late antiquity have been investigated to find out why we are so confused about sexuality. Peter Brown and Elaine Pagels (Princeton scholars who do not quote each other) have both sought the roots of our modern sexual disorders in Christian ideas of sixteen hundred years ago, without pausing to trace the route by which these ideas allegedly reached us. Pagels, in fact, has made a career of fighting her contemporary battles by thrashing ancient Christians she disapproves of. What would she do without western civilization?

On a more ambitious level, the Cornell sinologist Martin Bernal has worked out his own cultural identity crisis by writing multiple volumes of his *Black Athena,* on the "Afroasiatic" roots of classical civilization. His work, though flawed in a hundred ways, has found a ready audience among Afrocentrists seeking a different legend for civilization's past that wrests credit from the Greeks. That line of argument leads to some stunning abuses of fact by others, for which Bernal claims immunity from responsibility.

In so doing, he misses the point. His own first volume shows, with some biases, the process by which early modern scholarship invented classics as an explanation for the past (yoking Greek and Latin, excluding near eastern and African sources). But his, and our, failure of imagination is not to see the logical conclusion of this analysis: to forswear such legend-making in favor of a genuinely diverse reading of the present and of the many pasts on which the present draws. And so it is no surprise that his work has been an opportunity for those who would make other legends.

So, it is now reasonable to ask a classics professor, what value *does* the study of the western past retain? To answer that question we would first have to have a far better idea than I fear we do of what actually happens in school.

Schools and colleges construct curricula with an eye to content. "Teaching Shakespeare" is treated as an objective and invariant practice of high value. When my mother was given a copy of Shakespeare in about 1930 by a beloved eighth-grade teacher, she devoured it greedily, for she had never seen its like and could not imagine how better to spend her time. In a world of vastly various diversions, what does it do to students to compel them to counterfeit that experience? In a world in which autonomy and self-will are praised everywhere, curricular requirements are a very different thing from what they were in Stalky's time. At the very least, we need not think that the construction of a compulsory curriculum leads to the achievement of an education. At best a curriculum is a tool, and very often it is an obstacle.

One way through this dilemma has been limned by the literary critic Gerald Graff, in his *Beyond the Culture Wars: How Teaching the Conflicts Can Revitalize American Education.* He has an engaging idea, and one that doubtless works for some teachers: If traditional values are being debated, make the debate the object of teaching, read Afrocentrists alongside William Bennett, and discuss the issues. This approach keeps education firmly on the most conventional ground, then struggles to do unconventional things. The difficulty is that it requires us to assume that the students are interested in the conflicts, or have at least heard of them. This is liberal optimism of the worst sort. What happens pedagogically, moreover, is that what is lively and vital when the course is first worked up gradually ages and withers, until the teacher grumbles in the common room that William Bennett's *Book of Virtues* (or whatever) has gone out of print and he can't teach his course properly anymore.

The underlying insight of this strategy is that ours is already a culture permeated by irony. Skepticism about received messages is rampant, leaving any system that depends on transmitting those messages vulnerable. To use the space of the classroom to teach both the message and the critical reception and evaluation of the message is to create an opportunity to reach students at multiple levels. The best teachers have always done this; the challenge is to go on finding new ways to do so.

The approach I favor to teaching critical reception of the past might best be called "teaching the surprises." To the scholar whose work lies deep within the heartland of western civilization, the secret truth of the cultural traditions we inherit is that they are so diverse, polymorphous, and surprising. So my teaching strategy starts students where they think they are comfortable and then seeks to disorient and defamiliarize them so that they actually *look* at what they are studying. For a teacher who thinks such moments of epiphany vital to education, such a moment is a godsend.

But can western civilization supply an ample range of such possibilities?

Let me mention only two tales: one of monks and one of princes. On a rock a few miles from the coast of the west of Ireland, a place called Skellig Rocks and virtually the first landfall on the European side for all the storms that roll across the Atlantic on the Gulf Stream, some of the earliest Irish Christians evidently saw evidence of Michael the Archangel at work. (Michael was the angel of power who haunted high ground, and flashes of lightning doubtless evoked the power of his sword in conflict with the devil.) In homage to his power, a tiny handful of monks paddled out to the steep and craggy rock, made their way five hundred feet into the air along its cliffs, and built for themselves a tiny community: a handful of huts, a chapel, and some patches of carefully nurtured garden soil in which to plant. In so doing, they were heirs at some remove of western traditions of monasticism and Christianity, but to the modern eye they were frankly mad. Another hundred feet higher and at the other end of the rock, is a single tiny cell with a postage-stamp-sized garden plot—a hermitage for someone for whom the company of half a dozen fellow human beings on this storm-swept, bird-dropped rock where the only source of drinking water is rain, and the only source of food is sea-birds, fish, and a few legumes was too worldly, too social, too tempting, and who thus needed to get away from it all. In my teaching of the history of Christianity, very few students, except the most devout Catholics (one sees very few of those anymore), have any idea how to deal with the phenomenon of medieval monasticism. The most avowedly Christian of them, usually evangelical Protestants of one kind or another, are utterly baffled by a form of Christian life that bears no resemblance to anything they know in the Christianity with which they are familiar. Is that us? Is that our tradition? What part does that past play in the making of the beery and sentimental St. Patrick's day religion of the Irish-American today? What do we learn when we make the effort to link the two? Is that Christianity at all? Is it congruent with what Jesus said and did in Palestine? with what Renaissance Popes did in Rome? with what suburban churchgoers today practice? Finding the discontinuities—the surprises—makes clearer than anything else can the power of human community-making implicit in our ability to see and identify continuities over time. The idea of Christianity is clearly functional and powerful, whatever students think of it; but what *is* it?

At the other end of Europe and the other end of the middle ages, what do we make of the history of Matthias Corvinus Hunyadi? King of Hungary (d. 1490), he was an undoubted hero in the line of Renaissance princes. Grandly successful in battle, shaper of a united and independent Hungary, he was also one of the most dramatically productive of literary patrons. His library was the marvel of Europe for a shimmering collection of illuminated manuscripts that take the breath away. Now scattered to the four winds, even in isolation they demonstrate immediately the artistic and cultural level of his court.

But Matthias Corvinus was forgotten. He was unlucky in his successors and they were unlucky in their world. Already in his lifetime, the pressure of the Ottoman Turks was felt from the southeast, and after his death, the middle Danube became the battleground between Christianity and Islam, to Hungary's lasting disadvantage. The territory he had united was carved up again and ended in the hands of Hapsburgs from Vienna and Turks from Istanbul, and Hungary did not become the dominant force in central Europe that Corvinus' contemporaries might reasonably have expected it to. In addition, he is a challenge to study for westerners; there is no satisfactory book about him in English, and the difficulties of writing one are multiplied by the variety of languages (at a minimum, Latin, German, Italian, Magyar, Serbo-Croatian, and Turkish) required to deal with the sources.

Corvinus' story teaches a lot about the making of Europe, and reminds us that you could be a great Renaissance prince and not make it into the history books, through no fault of your own. We praise other Renaissance princes forgetful that their splendor is mediated to us by their worldly, military success and that of their descendants. We forget that linear narratives sketching one success after another do a very poor job of reminding us of the risks and the diversities of human experience. This surprising fact enriches our sense of history, while at the same time usefully reminding us of the limits of history.

We seem to have come a long way from Kipling and Housman, and from considering the place of classical scholarship in a few western societies. But the divagation is deliberate and necessary. Our educational institutions embody our most pretentious self-images. When we use them to teach "content," to impart "ideas," I fear we do our students little good. We habituate them to the myths of the tribe, and perhaps we impart some low-level serviceable skills, but we scarcely actuate their intellects. The study of the Greek and Latin classics and the study of the western civilization in which they are granted a leading role is valuable not if it produces the student who knows "the kings of England and can quote the fights historical, from Marathon to Waterloo in order categorical," like the Major-General in *Pirates of Penzance.* It is valuable if it gives a frame of reference, in some ways recognizable and in others alien and defamiliarizing, within which to see the present. Education happens if and only if that experience strikes sparks, shows students the unexpected, and energizes. Education is a discipline of *seeing,* seeing past resemblances to differences, recognizing the otherness of even the familiar, what Simone Weil spoke of beautifully in her essay on "attention" in school studies. If we lead students to do that, whatever the materials, we are doing education. If we accomplish it with the Greek and Latin classics, it is because we have a long tradition of doing that job superbly well, and there is merit in tradition that way. But we teachers do not automatically deserve a future. We must earn it by the skill with which we disorient our students, energize them, and inculcate in them a taste for the hard disciplines of seeing and thinking.

Chapter Six

AUGUSTINE TODAY: LINEAR NARRATIVES AND MULTIPLE PATHWAYS

Academic convention decrees that if you have spent years of your life studying a subject, you will be a suitable reviewer for books that weren't really meant for you to read. I have written at some length on St. Augustine, and so I am offered, and too often accept, books about him to review, even books addressed to an audience considerably less familiar with the subject than I. I think I am expected to be able to say whether the book is well-constructed for its purpose, but what that assumption neglects is that I am likely to be bored while reading at least part of the book. I used to take this boredom as a sign of my own importance and deftness as a scholar: if I didn't know so much, I wouldn't be so bored. I worry about that more now, and wonder just what these books have to say to someone who doesn't already know how the story turns out.

I find myself restlessly trying to get beyond the placid competence in matters Augustinian that I aspire to and demand of others. I want to see more clearly what Augustine has become today, and then to think a bit about where he might be going. While the value of this exercise to Augustinian scholars is obvious, those few of us with those skills and interests must recognize that our (and his) value to the societies of today cannot be taken for granted and must not be assumed to be fixed, stable, and unchanging. Such value is relative, and depends on the recipient as well as the source.

To examine the modern cultural horizon for Augustine's presence is to get a variety of signals. Bob Dylan's *I Dreamed I Saw St. Augustine* pays homage to the old wobbly anthem about Joe Hill (itself a ballad of a kind of resurrection), and Sting as recently as 1993 had a minor hit with *St. Augustine in Hell,* a pastiche of impressions and echoes ("The less I need the more I get / Make me chaste but not just yet" —the second line an ironic Augustinian depiction of his attitude when still young and unconverted). When I attended the annual St. Augustine Lecture at Villanova University a decade or so ago, I found myself sitting among students who had been press-ganged into attendance by their teachers. They were trying to remember who Augustine had been, and the best informed among them reported that Augustine had been "the patron saint of, like, criminals" because when young had done "everything." They seemed unconcerned by their ignorance, and I was only moderately relieved when some minutes later it finally dawned on one of them to connect the saint to the Order of Saint Augustine whose priests founded and preside over Villanova.

Augustine's reputation of early profligacy is firmly attached to him. (Indeed it drove me to buy my own first copy of his *Confessions* somewhere deep in high school, when I thought I would find it both improving and titillating. For the nonce it proved to be neither, and if it has now become at least somewhat improving for me, it still fails to titillate.) The power of that reputation is linked to Augustine's other widely held reputation, as putatively hostile toward human sexuality. Whatever the facts of the case, he is required by our culture to play the role of the youthful libertine turned into the elderly oppressor of human sexual feeling. That is the irreducible, if misleading, image that he carries with him today.

Still other modern issues bubble to the surface when his name is mentioned. Some readers of the *Confessions* have found reason to believe that it veils admission of practices that would now be thought homosexual. Their argument would be more forceful if they could agree on which passage of the text gives that hint. Until they do, a fair judgment must be left in suspension. Meanwhile, we can contemplate the intensity of feeling that many bring to the attempt to force its resolution one way or the other.

Similar intensity of feeling in another area again tells us about ourselves more than him. His African birthplace inspires some to imagine and others to assert that he must have been "black." The evidence is richly ambiguous. In the ancient Mediterranean world, the standard against which other skin colors were measured was swarthy and sunburned, with disdain shown for the intellectual failings and rude physical prowess of blonde, fair northerners, who were better known than the darker denizens of central Africa. Augustine knows of such blacker Africans, but his allusions to them tell us very little about his own color, ethnicity, or the like. Readings and readers of Augustine who pursue these questions may indeed be using him quite legitimately to learn more about themselves and the preoccupations (sexual and ethnic, for example) of our time, but the most valuable lessons of the past are the ones we don't go looking for and least expect.

The one place where Augustine's *doctrine* is continually invoked as though it had authority is in the matter of bellicosity. Whenever American troops are sent in harm's way, it is reasonable to expect the margins of *Time* magazine to sprout a thumbnail-sized Old Master representation of Augustine, accompanied by a potted summary of his doctrine of the "just war." In the last such case, George Bush fervently invoked Augustine's authority to justify the Gulf War (though with some unfamiliarity, for the saint came out jumbled up as "Saint Ambrose Augustine").

The debate can reasonably go on elsewhere to just what extent the doctrines ascribed to Augustine on those occasions represent him fairly, but for our present purpose what is important is the way the general authority of the cultural ancestor is suddenly made very specific and very powerful when a current applicability is found, even though in many other areas the identical authority is rejected, neglected, or simply unknown.

The Augustines the modern world remembers do not much resemble the Augustine of the encyclopedias, nor the Augustine of the books I review. There he is a solemn figure of high intellectual power, a man of words and arguments, defined by his doctrines. The *Great Books of the Western World* series devotes a whole volume to him, then leaps forward eight centuries to find its next worthy writer in Aquinas. No history of western thought or the like is complete without him. He is both Christian and Platonist at the same time, invoking two other labels that mask far more difference (it can be argued) than they supply unity. (Never mind that neither label means quite what it used to, nor that the Platonist label in particular is largely obsolete. It is what Augustine is, and if he seems a bit less urgently present than he once might have, it is perhaps a testimony to the obsolescence of another generation's labels.)

A little outside the somber handbooks and histories, Augustine's image begins to change. The doctrine is not always the man: the life is the man. Augustine the man is there for us in the *Confessions* of course, but now we flock to his text to psychoanalyze him. This is not the place to assess the validity of the methods used. The point to observe is that we think it important that this man's life, even though he is long dead and lived in a remote land far from centers of power, both can be and should be subjected to such scrutiny.

Consider, for example, the line of descent that runs from the 1930s to the 1990s, from a popularizing biography of Augustine by Rebecca West through Elaine Pagels's *Adam, Eve, and the Serpent,* to several studies by Harvard theologian Margaret Miles, most recently her *Desire and Delight.* It is not farfetched to see continuity in their common preoccupation with an Augustine as father figure, who resembles no one so much as his own father, quick to anger, brutal in deed or at least word, marked by an almost frenzied authoritarianism. In that view it is, not important whether the cruelty resides in the author or in the scenarios of punishment and hell that he sponsors. The effect is the same. This brutal and domineering Augustine is a modern discovery, and whatever we make of his role in the history of western thought, this Augustine is not what he has been. Rather, he has remained what he was, as the rest of the culture has gotten in touch with its feelings and gentled down. The poignancy of this judgment we want to pass on him is that it condemns a role that he chose and enacted in part because it was the necessary role in his culture for assertion of any self at all in his class.

So what is the remedy for our inclination to judge the past by the values of the present? It would seem that one could go to the sources. If we do for Augustine, we are repelled by five million surviving words in Latin, not all ever translated into English, and few enough of them readily accessible wherever a reader might happen to be. The natural and most common recourse in the English-speaking world is to retreat to the life of the man again, either to the *Confessions* directly or to distinguished modern scholarship.

But for one who has studied the *Confessions* as intensively as I have, to listen hard to the way the *Confessions* are read is to see again selectivity and partiality. A pear theft, a moment of conversion in the garden, Augustine's relations with his mother up to and including a moment of mystic vision at Ostia—these are the backbone of the *Confessions* that my students, friends and colleagues have read. The last books, with their extensive not to say obsessive exegesis of the first chapter of Genesis, have been ignored almost completely. (In some older modern translations, they were even simply omitted as of no interest.) In an odd way these few episodes connect to the armature of the narrative quite faithfully, but they add up to a book with almost all the artistry and effort the author put into it removed. There is an almost universal truth here, that books whose power is greatest succeed in staying in the cultural memory only fragmentarily and partially. Think how little we remember of *Hamlet* or *Huckleberry Finn*. The art does its work, but we remember only the most pronounced traces—a caricature.

So who is the Augustine we remember when we remember the *Confessions* as a caricature? We think we are remembering the young Augustine, whom we romanticize and religiosify according to our inclinations. Invisible to us as we do this is the shaping influence of the middle-aged Augustine, bishop of a provincial city with an agenda to work out as he writes the received version of his younger self we've come to know as *Confessions.* Just how far and how deeply Augustine was "converted" is a point over which we should linger more than is our custom. We should be wary of being so far bowled over by the narrative of that middle-aged churchman that we believe a too-transparent and easy story of fall and rise, sin and redemption. We think we know the story of Augustine, when what we know is the story that Augustine would tell.

In principle, the ambiguities of any kind of presentation of self or other are known, and even found enticing by many. Literary critics from Kenneth Burke to Eugene Vance to Jacques Derrida have found in this text a tissue of images, ideas, and artistry to beguile the most postmodern of imaginations. The feebleness of the word to control reality, the fiendishness of the struggles with which an artist like Augustine uses the word to find a semblance of that control, and the ultimate failure of those struggles—all these are modern themes that resonate through Augustine's text. Most recently, Derrida, himself once a boy growing up on the *rue Saint Augustin* in Algiers, reflected on his own life, culminating in the death of his powerful mother, in a work he calls "Circumfessions," with a double allusion to Augustine's liminal prayer, "Circumcise my lips" (*Confessions* 11.2.3), and to the *Confessions* themselves. But though we enact the words of the poststructuralist and postmodern theoreticians in our lives, we deny them when pressed, ignore them when we can, and the Augustine who *might* emerge from a consistent application of these lines of reasoning so far shows little sign of dethroning the Augustine we have already created out of his *Confessions.*

But the serious reader of Augustine in our generation has not often stopped with the *Confessions,* at the same time rarely continuing on to read the other five million words of Augustine's surviving works in order to form a more perfect picture of his achievement. Instead, since 1967, such a reader has turned aside early to read the luminous and fluent biography written by Peter Brown. No one interested in Augustine should delay for a moment reading this book, if only in order to join the community of those who have read it. But no one who reads it should refrain from thinking about the implications of that reading.

First, the canonizing act of biography in the modern critical sense waited long to come to Augustine. As late as 1963, the best new book in that vein was Gerald Bonner's *St. Augustine: Life and Controversies.* The title reveals two preoccupations: Augustine seen as saint, and Augustine seen as warrior in ecclesiastical controversy. Bonner's book is still in print and still extremely useful, but it is Brown's book that has taken the world by storm. Some of this is owed simply to the talent and the prose of the author, but more is owed, I believe, to its devout conformity to the norms of biography as we practice it now and its cunning adaptation of Augustine's own autobiographical inclinations to its purposes.

Like a popular modern biography, Brown's book is implicitly psychological on every page, using categories clearly situated in a post-Freudian tradition, but equally importantly, it denies any attempt to engage in psychoanalysis. Thus it avoids being pigeonholed as psychobiography, but usurps most of the benefits of such an approach. Further, since Augustine's *Confessions* do provide a skeleton narrative of his life up to age thirty-three, for nearly half the book, the willing reader is hustled along through familiar, or at least readily verifiable, territory, with scholarship enriching the story but not fundamentally changing it (this is interesting and important) on any point. By the time Brown's narrative leaves the *Confessions* behind, it has a full head of steam and the complicity of the reader.

One implicit claim of biography as a genre is that it reduces a whole life and all its complelxity to a single narrative line. What that claim elides is a fact deeply rooted in the fact of authorship: Every author of biography can write his or her own confessions ("how I got here") and no author of biography can write from experience a whole autobiography—for the simple reason that no such author writes from beyond the grave. Biographies thus fade badly toward the end. The romantic self we imagine and describe in them can be traced through youth and maturation but not beyond.

Such biography becomes notably unsympathetic and even disoriented when its subject ages. The Romantic imagination succeeded in depicting normative patterns of youthful development, but it failed to carry such imagination forward through the vagaries of middle and old age. Nobody ages well in biography: Augustine in Brown's book is no exception. Brown's older Augustine has become a cliché of the scholarship since: as he grew older, he became dark and angry, hostile to human freedom, impregnable to argument or persuasion, and deeply suspicious of human sexuality in all its manifestations. That older Augustine glibly explains away the practices of some austere branches of modern Christianity, and thus he—along with Brown's book—has become a fixture on the landscape. It seems that all biographies of Augustine take for granted that he turned out exactly as one would expect, a stereotypical old clergyman hurling imprecations against youth, vigor, and sentiment.

To point to this practice is to question it without replacing it, and it is important to see that, as traditionally practiced, scholarship cannot replace it. The conclusions are implicit in the biographical and even historical practice; implicit again in the expectation that scholarship is an individualist practice, validated by being made public, but in the end depending on the solo scholar working alone. In theory, we believe in collaboration and interdisciplinary work, but in practice, we receive still with greatest pleasure the achievements of the individual virtuoso.

And that habit conditions us as well to expect the single persuasive narrative line, something we should know to be absurd.

For consider the life of any contemporary, famous or obscure: How far is it possible to achieve common consent about a single narrative line to construct such a life? Should we remember that our friends and enemies always disagree about us, and with reason? But we resist these observations because of essentialist ideas: we must be *something,* some one thing, and in principle we want to think that scholarship or psychology or detective work can lead us to that unity.

But all the evidence is against us. People are too protean in their relationships to be reduced to a single gestalt, and biography is too linear to do them justice. What I am suggesting is that in Augustinian studies, but by implication in others as well, the traditions of scholarship tell us that a single truth is to be found and condition us to ignore our visceral experience that the single truth is always found masked by multiple perspectives and arguments.

At a moment when all the conditions of scholarly discourse are about to be upended by the transformations of electronic technologies, there is an important opportunity to reconsider what it is we scholars seek to do, how we seek to do it, and what we can reasonably expect to achieve. What follows sketches what I believe we can expect to do with Augustine, and by implication, what we can expect of traditional humanistic scholarship in an electronic age.

The central fact of our future is diversity. The single-author, linear-structure monograph will survive for a while, but it will very rapidly become in fact what it already is in principle: a component of a larger whole. Online publication of monographs will facilitate a multiplication of approaches and comparative interaction. Just as the Eusebian canon tables began the process of reshaping the four Gospels into a single database, so the online Augustine takes the library shelf to the next level of usefulness and interaction by making juxtaposition and comparison easier and more natural.

Second, primary and secondary materials will interact more powerfully than before as both are online side by side. Scholarly discussions will quote the original by pointing to it, and leave the reader to explore the original context, not just the few words or sentences most apposite. Conversely, texts will acquire structured commentaries not by single hands but organized out of the work of many. A controverted passage of the *Confessions* online can be linked to multiple articles and treatments directly, and then also to intermediate links that would seek to organize and arrange the body of secondary literature. The "variorum edition" is a print phenomenon that has never been widely popular. Its time may well be coming soon as it is possible to directly link texts to a wide variety of scholarly discussions.

This is not to say that the reader's task will become easier. If anything, the burden of responsibility will seem crushing to many. The great risk will be that providers of shortcuts will find devious and enticing ways to flourish and seduce. When you are at sea in a mass of information that challenges you to think and judge for yourself, the salesman will certainly be readily at hand to offer easy surcease. There will be a real struggle for control of the life of the mind in cyberspace. And sociology and economics will become players in that struggle as they help us determine how much time and effort Augustine is worth and who will spend years of preparation in order to use such resources well. The shift from an economics of scarcity to an economics of abundance becomes painfully relevant and threatens to change the landscape dramatically.

Nevertheless, the migration of information into cyberspace will not proceed in any comprehensive, predictable pattern. The prestige and influence of specific ideas and schools can be managed relatively easily in a known environment of publication and communication. Transfer them to a new environment, and those who get there first get to define the space. Received and established opinions have no particular advantage. There might well even be a negative correlation between prestige and promptness. The most established academic ideas may feel the need less urgently, while innovation, curiosity, and chutzpah will lead the unorthodox to take advantage of a new environment more readily.

So it is that the veteran Augustinian looks at cyberspace and inevitably begins to think about how to organize Augustine anew. The old categories of dogmatic theology and conventional biography don't compel here. Suppose we started over. Instead of linear narrative, what if we look for multiple pathways and links?

One path would let you follow Augustine's own intellectual agenda through his career. In one of his earliest works, he says that the only things he wants to know are "God and the soul, nothing else." In the world of print, it is possible to imagine a monograph, perhaps a weighty French thesis, that would take that program and pursue it through all Augustine's works, reading them as explorations of just those two themes. This would be intellectual biography where the subject sets the agenda. The natural and justified criticism of such an exercise would be that it would have some difficulty sustaining the proposition that this is a privileged way to read Augustine. The author would be challenged to show that this preoccupation with God and the soul was real and determinative throughout Augustine's career.

But cyberspace is a place where even the serious engage in play without penalty. If a fullish database of texts is online, it begins to be possible to string together, tentatively and one piece at a time, a line of interpretation deliberately presented as experiment. The experiment can begin to be "published" at an early stage as comment is invited, and it can grow and build in public. If its value does not demonstrate itself, presumably the project falls apart or changes direction; if it receives encouragement, it goes on. This is not so different from the present practice of presenting articles publicly for comment on the way to a larger project, but this method can reach a larger audience at a more preliminary stage of the work. It presents itself to an audience that can immediately grapple directly with the texts and test, improve, refine, or criticize what is under way. Instead of publication that says "This is how it is," we have a form of public performance of scholarship that *asks* "What if it were this way?" Publication of this sort becomes a form of continuing seminar, and the performance is interactive, dialogic, and self-correcting.

An online Augustine creates a space that belongs more nearly to Augustine, that facilitates navigation more powerfully than any print archive can do, and that encourages systematic and comprehensive questions that generate results from the whole range of a huge oeuvre. The reasonable expectation is that the isolation of platonic or any other elements, like radioactive isotopes, will seem less and less reasonable a course of scholarly action, and instead we will be challenged to look at the tenor and context of Augustine's writings. It will be less possible to separate off a single line of thought. Inquiries that connect and integrate will enrich a common resource and our sense of possibilities. The results will be more resistant to distortion because the results will still be located in the space of Augustine, not torn off and taken to another space.

At the same time, the nonspecialist is enfranchised as well. It is a truism of many fields of study that existing categories have a blinkering effect. We all speak with praise of interdisciplinary studies and are frustrated that we cannot enact them more effectively. Social historians, for example, struggle to get into Augustine's territory, but are blocked by a variety of forces, not least the sequestration of his books in traditional disciplinary categories. At my university, copies of Augustine's books in Latin may be found as far away from the ancient history seminar and reference collection as it is possible to get while remaining in the same building.

By these means the cult of personality is reinforced. To study Augustine is to study "Augustine." We have been historically unable to see him and his texts without his reputation and without his context—that is, without the context that moderns have chosen to give him. Here the disorienting effect of the move to cyberspace has a strong positive effect. Gilbert Chesterton wrote of wanting to write a story of a man who got in a boat and sailed away from England, coming finally to a strange land where he had many wonderful adventures until he discovered that by a quirk of navigation the strange land was England itself—the strangeness and wonder were made possible because he did not know where he was. Traditional scholarship offers few such opportunities. The heuristic quality of life in cyberspace and the ease with which multiple paths can be created will let us create such opportunities with ease and indulge in the high-spirited play of manipulating the tokens of the past in as many different ways as we can imagine.

The Augustine we can reconstruct through cyberspace navigation is the ordinary Augustine, the Augustine who was a minor celebrity in his time but mainly the bishop and judge and orator of his community. On one memorable day in 418, Augustine went to a neighboring town and quelled a riot with his oratory. When I think too much about it being "Augustine" who did this, I'm not surprised. Of course Augustine was a powerful figure, but talking down a riot is no mean feat. It is far from self-evident that he could do such a thing.*

Once we begin to think in this multipath way, we find the fundamental challenge that the new media will pose to the world we have chosen to live in. Richard Lanham (professor of Renaissance literature turned cybertheorist) is eloquent about the austerity of decorum that print texts have achieved, and the departure this represents from the efflorescence of image and adornment in late medieval books. It has certainly been a profound quality of the great age of the mass printed book (by which I mean the nineteenth century and its afterlife) that formal public life would be dignified to within an inch of its life. Our serious public dress was dark and self-effacing, our public demeanor stiff and quiet, and the division between places and times for work and for play rigid in the extreme.

Just to say that is to make the contemporary reader realize how that hegemony of the solemn is breaking down. Sometimes it makes us nervous that our grade school classrooms are not quite so stiff and uncomfortable as they used to be, and we hanker for a bit more discipline, but no one seriously seeks to go back to the decorum of two generations ago. Look for a moment at old film of baseball from the glory days of the 1920s. Watch Babe Ruth hit a home run and see twenty thousand adult males wearing dark suits *and hats* leap to their

*"To see what has become self-evident as something that was not originally self-evident is the task of all historical reflection" (Hans Blumenberg, *The Legitimacy of the Modern Age* [Cambridge, Mass., 1983] 594).

feet to cheer. It is a preposterous scene. Go to a presidential campaign rally today and see how much of the trappings are those of a rock concert. We have gone to the opposite extreme, and somehow it makes sense.

But is the high dignification of western civilization about to seem similarly preposterous across the board? At the very least, we can put back on the table an issue usually seen only from the dignified point of view (as in Norbert Elias' famous *History of Manners*): what is the place in the history of our society of the specific form of social control implicit in our norms of decorum? How did they come about? How necessary will we think they are?

This issue crosses traditional study of the life and work of Augustine in a surprising way. For in a traditional reading of the past, a dignified figure like Augustine seems quite natural to us. What we have so far failed to make a part of our consciousness is the history of the dignification of Christianity. Conventional pro- and anti-Christian reading of Augustine's fourth century emphasizes growth and enhancement of power in remarkably similar terms. What at this remove is surprising is the quality that both sides in such partisan wrangles take for granted, the solemnification of the religion. In some respects, the process had been under way before Augustine's time, but surely no religion that started with Jesus and his renegade Jews could be assumed to develop as hieratically as Christianity did. Christianity was the high-tech religion of late antiquity, using the written word resourcefully to create and shape itself.

But in that power was also a trap that Christianity did not escape. The religion we know was materially shaped in that fourth century when it became the official religion of the Roman world. In that period we see Christianity settling decisively for the textual decorum of Roman literature, and Christian bishops taking on the power, the perquisites, and the demeanor of the Roman public servant. The intersection of public Rome and textual Rome in the creation of the Christian virtual library made a Christianity defined by publicity and textuality: in other words, by a process of dignification that eventually became determinative for the whole of society.

Such decorum can scarcely be sustained in public and private life much longer. The play-like character of the electronic environment will compel us to address a question about our past that we have not done a good job of perceiving, much less pursuing, before. The Augustine who chose Cicero and Ambrose as his models found a way to reconcile the religion of his mother with the culture of his father, to turn the Bible into another official text, and even to turn Christianity into another version of Rome (this is the nonobvious way to read his *City of God,* but an important one I think).

And behind that dignification is the discovery of another fact about our past that comes into question in this age as well. In the year Augustine became a priest, 391 C.E., at age thirty-six, he wrote a book called *True Religion.* The modern student of Augustine will have read that book and will, if at all like me, have taken it for granted. Of course Augustine would write about true religion—it's what he stood for. But if you read the text closely, a radical innovation, not made by Augustine but clearly expressed by him there, emerges. He criticizes Roman religion for its disconnection between religious practice and truth. For him, the Roman philosophers, who speculated about the nature of divinity in high abstract terms, then participated in cult activity on other terms, were hypocrites. For Augustine it has *become* obvious that what you do about God must be coordinated with what you say and think. His cult behavior—a thing of festivals and magic—is coming under the control of texts. And when that happens, the criterion of truth gradually changes.

There is no western language in which the word for "true" or "truth" is a native and obvious one. In one form or another words like "true" and *vrai* and *wahr* are words about the reality of things, not the verifiability of propositions. But the practice of the written word gradually shifts the locus of truth from the individual to the page. No longer is evaluation based on the reliability of the speaker, but on external manipulation of words on a page. Truth is independent of the speaker and in that way external to human beings. It becomes objective and powerful. We are more skeptical today than people were a hundred years ago about the ways human beings can manipulate textual truth to their advantage, and we must acknowledge that textual truth, on the other hand, has made possible much of what is distinctive about our society. The advance of science as a collaborative activity depends on that external truth being open to inspection and refinement.

But we need to see these practices in perspective. For one thing, the decorum of those textual practices, ubiquitously distributed and gradually permeating all aspects of society, does impose control of various kinds where it might least be expected. Michel Foucault is droll in a deadpan way when he argues in the first volume of his history of sexuality that the decisive controlling moment in the history of sexuality was the creation of the counter-Reformation confessors' manual, full of details of prescribed and proscribed bedroom conduct. He argues that this represents the stage at which even the bedroom is textualized beyond spontaneity, and he gives his point a wicked twist by arguing that the twentieth-century sex manual represents no essential difference, just a change in value polarities. The user of such a manual, no less than the priest with his confessors' manual, takes the view that what goes on in a bedroom can and should be determined by the prescriptions of authorities outside. That marks a dramatic difference over the very different medieval practices guided by, in theory, the same moral precepts as prevailed in the post-Reformation churches, and offers a model for the transition I identify here.

The value of this observation is twofold: first, it throws light on our pasts and thus on our present, but second, it gives caution for the future. Our science fiction is written in a binary way. Either we will continue with an accelerating curve of scientific discovery and rational progression or we will subside into ignorance and barbarism through some failure of will. What just now begins to be visible is a third possibility going in quite a different direction. In this view, the advance of science will be overtaken by the advance of play. An economy of amusement will replace the economy of material sustenance, and the progression of humankind will be in directions that our nineteenth-century progenitors would regard as frivolity.

The electronic environment has the power to defeat dignification: can we bear the results? We must not idealize them as the triumph of fresh human spontaneity. In the best case, they would be themselves no less artificial and constructed than what we leave behind. We will only discover as we go along whether such an environment can sustain itself materially and transcend the preoccupation with such sustenance. In a world where material sustenance is a critical question still in many lands, the advance of some societies to another kind of economy may impose acute political stresses with disruptive and destructive results. There is no linear progress in history, but halting progressions on many fronts, and prediction is impossible. Some qualities of the future may be surmised, but how those qualities will interact is impossible to realize—and that uncertainty leaves for us our mission.

I have deliberately suggested the function, possibility, and impact of humanistic discourse in the present age. One cannot read Augustine or other long-dead worthies, I believe, without the most acute sensitivity to the realities of our own age and all the links of imagination and tradition that bind us to our pasts. On the other hand, it is enriching and suggestive to engage in that reading because of the way the study of the past engages so fully our commitments and our opportunities in the present. To let the past be the past, without making it into an allegory about ourselves, is to learn best from it what we need to know for the present.

Chapter Seven

THE NEW LIBERAL ARTS: TEACHING IN THE POSTMODERN WORLD

Whenever I go to a zoo, I'm a bit excited and unsettled until I come to the tigers. I want to make sure that there are at least a couple of them there, and I want to reassure myself that they are in good spirits. To me, tigers represent the best of the animal kingdom: extraordinary beauty in fluid motion that masks a power that would gladly snap my neck in an instant, utterly disdainful of the dominion humankind has seized over other species and of the threat of extinction itself. I am shaken and thrilled when I see the tigers, and go away reassured by the beauty and the danger of the world we live in.

I chose a career in scholarship and teaching in the same spirit. There is no refuge from reality in teaching, no orderly life in a kind of Disney World of the mind where nothing really dangerous ever happens and a predictable good time is had by all. School often presents itself to the imagination as that kind of sanitized theme park, but as school becomes university, risks need to be taken. Failure to see this is one of the root causes of our so-called culture wars, where both left and right argue over how best to manage the Disney World University and which exhibits to put on display. In the face of such a spectacle, moderate and practical people turn away, telling their children to study accounting or some other less embattled subject with the result that the humanities are often held in low esteem.

We've been there before. Perhaps we've been there ever since Nietzsche, the prodigiously young professor of Greek and Latin classics, began jotting the fragmentary notes that come down to us among his *Unzeitgemässe Betrachtungen* (Observations Out of Step with the Times) as "Wir Philologen" (We Philologists or, less jarringly, We Scholars). In these notes he lambasts the routinization of the scholarly charisma in his own day with the telling and true point that the techniques of humanism had evolved as weapons of rebellion in the service of human freedom, but had become instead the pedantic tools of comfortable bourgeois professors seeking to preserve a social and cultural order.

Nietzsche never finished those observations, and in the years after his professorship he followed a vision of idolized Greeks through his own private Disney World to culminate in the autobiographical *Ecce Homo* with its chapters on "Why I Am So Clever" and "Why I Write Such Good Books." In his case, the failure of the traditional arts led to a personal dead end.

In an odd way, Nietzsche was partially reenacting a personal drama from fifteen centuries earlier. Around 400 C.E., another precocious professor of the classics threw over his career in his early thirties as the mere "hawking of words" and struggled to express a new cultural model. After almost ten years, Augustine too could scratch out only an incomplete, fragmentary, and too-theoretical exposition of what he was reaching for (a handbook for teachers), and he broke off that exercise to write instead of himself. What he found when he turned the spyglass on himself was a being opaque to its own scrutiny and intelligible only *sub specie aeternitatis.* Having written his *Confessions,* Augustine went on to produce a flood of texts whose power depend in no small part on their virtuoso balance of authority and humility.

It is no accident that the aptest *comparandum* for Nietzsche's revolt against the educational system comes from the late antique Mediterranean. The totemic notion of liberal arts that we recreate in our modern institutions took decisive shape in the fourth and fifth centuries, as an amalgam of old school practices and new philosophical ideas. Augustine, heir and reformulator of that liberal arts tradition, is of no small interest in this regard.

The idea of a coherent cycle of liberal arts is probably best attributed to the Roman polymath Varro of the first century B.C.E., but we are imperfectly informed about his ideas and practices and cannot agree today whether he offered only something pragmatic (the majority view) or something rather more ideologically venturesome (the minority view, to which I subscribe). But in the fourth century, a fresh ideological charge is first discerned clearly and the fourth through sixth centuries saw the doctrines and their accompanying practices take shape in the hands of writers like the Africans Augustine and Martianus Capella and the Italians Boethius and Cassiodorus.

For that world, the liberal arts were not meant to fit you for life in the workaday world, nor to make you a good prospective citizen. Their aim was philosophical, even mystical. The world of appearances and material being was full of distractions and confusion, but the liberal arts, offering mastery of language and number, would disaffect the mind from the charms of this world and lead it to ascend by graduated steps through this world toward that which lay above. The multiple liberal arts all led to the same goal: in neo-Platonic terms, "The One";

in Christian terms, "God"; in all cases, a fundamental metaphysical unity that animates the totality of things.

This belief in a unified totality explains one of the oddest if most obvious things about our universities of today, namely their blithe disregard of the rule of economies of scale. A major university (to say nothing of most minor universities and small colleges) still insists on diversifying itself across the disciplines, offering courses in everything from biology to medieval art, no matter how few students want to pursue them. And yet, the institutions that still pursue these many disciplines are not quite *uni*versities any longer. (Clark Kerr's word for them, multiversity, is apt even if we can't quite accept it in daily use.) Even where the core curriculum has been shoehorned into an institution, there is today no organic relation among the parts of the university.

Instead, a university resembles a shopping mall, where vaguely unrelated stores come and go with the whims of fashion, and management's job is to make sure there aren't too many shoe shops (nor too few) and that they are all economically self-sustaining. In this world, students come and go from various courses and departments without being asked to think about the whole, without being challenged to find the missing pieces. Our institutional vision has failed when it is not clear to our deans or our students what the classics have to do with sociology, or clear to anyone what the liberal arts have to offer the businessman on a flight to Japan other than a badge of class distinction, distracting entertainment, and a sense of cultural superiority.

As we contemplate the changes the future will bring to universities and how to adapt to them, we must reexamine what business we are in. If we need a monument to error in this reconsideration, we need only look around us. If the railroads of the 1950s had known they were in the transportation business instead of the railroad business, the joke goes, more of them would still be in business. Similarly, if we think we are in the youth camp business or the fifty-minute lecture business, we may still be in those businesses (some of us) forty years from now, but there won't be as many of us, the paint will be peeling from the walls, and the dormitories and lecture halls will be far quieter and more tranquil places.

This may seem like extreme alarmism, for the modern university is a huge presence very near the center of our economy. But we have seen in the last generation that wealth and power alone are no guarantees of survival. Acres of closed steel mills, whose furnaces once powered the national economy, tell us that. We are immensely fortunate that academics have been in the front line of computing and networks. This gives us now an advantage—technical, intellectual, and even just financial—that we would be fools to squander.

The changes this technology will bring raise a host of questions for academics: What will we do on the superhighway? What happens to higher education when every student has a link to a flood of words and images, metastasizing in every imaginable way from around the world, and when every teacher and every student can reach out to each other at all hours of the day and night? The short answer is that we don't know; we will soon, and are even now finding out; and in so doing we will reinvent pedagogy and the university as we know it now. Technology will do what it always does: provide tools. Those tools may eventually shape their owners, but they are always assuredly instruments with which their owners may pursue their own aims.

The most thoughtful exploration of the educational and scholarly possibilities before us to date is Richard Lanham's 1993 volume *The*

Electronic Word. In this book, Lanham draws on a career in the study of academic rhetoric supplemented by years of practice leading a great university's program in English composition. He makes a theoretical and dazzling case for the centrality to the electronic future of the very discipline to which he has given his life: rhetoric. The electronic media, he says, show us how to use our communicative skills self-consciously in an environment in which we do not seek to possess truth but to create it collectively. He justifies his claim with the observation that rhetoric is indeed a very old discipline and that a magically benign cultural continuity emerges from its reappearance.

Some will dispute that we live in a magically benign world. Lanham concentrates on the place of language in the world beyond the electronic screen, probably to excess, and one technologically venturesome and admirably humanistic dean of a school of engineering expostulates that from where he sits in the academic trenches, there is no less a crisis in our communities' attention to quantitative skills. By quantitative he does not mean only the formal study of mathematics, but something rather more vital to a healthy society: the ability to use quantitative data intelligently in the ordinary reasoning and discourse of human life.

What Lanham proposes, moreover, is familiar: a renovated curriculum. Not quite a core curriculum, but some core techniques. One has the feeling in his pages that a university that rose up to draft a new grand plan for its undergraduates with rhetoric at the center would delight Lanham and still leave his engagingly wry and ironic side suddenly apprehensive at the memory of having seen quite enough of curricular plans in his lifetime. Such plans require us to have a common vision in some underlying story we tell each other about ourselves, even if it is a story as fresh and engaging as Lanham's. But no such metanarrative has a reasonable chance of winning wide acclaim today. Only sects have metanarratives now, and sects prefer their myths to be extreme. Lanham is not extremist enough to have a sect of his own, but his prescription is too sectarian to be successful on its own terms.

With the failure of the metanarrative, we must imagine something else instead. At the outset, we should look clearly and frankly at what universities are, what we do, and what we can do. We must have no fear of cheapening ourselves by stooping from the heights we have sought to dwell on. We are, frankly, cheap enough already. Those of us entrenched are too comfortable, and have too many glib rationalizations for the inefficiencies of our teaching and the inequities of our professional hierarchies to hold ourselves up as paragons. If we know at the outset what we are and what we must do, then we remain as we are, dwindling like characters in Angus Wilson's *Anglo-Saxon Attitudes* into shabby self-parody and the determined show of gentility that an emigré nobility clings to. Some of our number already seem to prefer that life, and fairness insists that they be left to indulge it. They need not be emulated or encouraged.

One fact in this reevaluation is central. The technologies now in hand break down barriers, blur boundaries, and facilitate connections. Our task now is not to create a new Greater Disney World or define a future. It is rather to explore openings, multiply possibilities, and venture down enticing pathways. It is too early for grand plans and instead a time for exuberance. We will pay the price for that exuberance in due time, but if we stint on the exuberance now, we will be able to afford even less in the future.

As we use technology to link the verbal and the quantitative in new ways, we realize that this has always been a possibility, but the distances have deterred us. Lonely explorers have ventured out to make contact, but our traditional disciplines ride heavy and low in the water, while our campuses are broad, often physically divided between humanistic space and scientific space. A virtual campus need not respect those divisions and can make the distances easier to cross. One area in which this has already been happening for years, quite invisibly, and quite without anyone thinking deliberately about how to make a liberal art central to the institution is in the study of "artificial intelligence." Philosophy, linguistics, computer science, psychology, and sometimes even medical faculty have a common interest in cognitive science and in the possibilities of "artificial intelligence," but the curriculum never knows quite where to put courses that might be taught in this vein, and the faculty involved have to work against the institutional grain to find space for collaboration.

Cyberspace offers a more flexible version of institutional reality in which we could find a way for such work to be very central indeed. That flexibility lets us imagine an institutional reality that need not be cast in stone quite as much as our existing campuses are. Departments are useful disciplinary alignments, but are hardly the only way to conceive the intellectual relations of a faculty. Deans and provosts speak wistfully of making departmental barriers less impermeable and seeking more flexible organization, but they rarely venture very far. Cyberspace is the place where people whose offices are several blocks from each other can be encouraged to shape new alliances in a parallel virtual university, to experiment with the intellectual and pedagogical practicality of new affiliations that can always be given bureaucratic reality after the fact.

Of course the deepest barrier on our campuses separates the liberal arts from the preprofessional schools. No one ever thinks of law or business or accounting or dentistry as part of the liberal arts. Those who know from the beginning of their university careers that their ambitions lie in these directions are privileged second-class citizens, rather like the *equites* (knights) of old Rome—of lower social standing than the senatorial class, but for the most part quite a bit wealthier. The preprofessional student has the prospect of economic success and the brash American prestige of money, while students in the arts and sciences are encouraged by their elders and betters to accept a lower economic prospect in return for a more venerable, but perhaps now more threadbare prestige. Venturesome and imaginative humanists will find a way to bridge this gap.

Some things we already do in many institutions, heuristically, chaotically, opportunistically without any concomitant theorizing. The tools are already in hand to make transformative change—and I would not have said that as recently as 1993. We can make some good surmises about technologies that are coming to help us further, but even if we have only the PC and the Internet of 1998, we have enough to revolutionize education in startling ways.

At the University of Pennsylvania, multimedia textbooks, World Wide Web online teaching resources, interactive email between faculty and students and between the university and the rest of the world, and even such apparent exotica as MOO online conferencing software—all these tools have already become everyday practice for many of us, and we are hard at work exploiting their power and trying to make sure their power does not overwhelm us. Even where faculty have only started to struggle to use the new technologies, the librarians, swiftly shedding any remnants of a dowdy image from former times, are already creating balanced information centers that lead the struggle to find a way for print and pixel to coexist for as long as we need and cherish both.

Another approach, resource-based learning, is a buzz phrase, but it points to a powerful trend. We can create teaching tools interactive enough and rich enough to let students seek them out and work with them at their own pace. Such tools do not directly address the central educational mission of motivation and direction, but instruction that is available when the student needs it and powerful enough to sweep the student along can reinforce motivation and accelerate progress. This can be done most obviously for content-based instruction in specific disciplines at a fairly elementary to intermediate level (whether to replace traditional courses or to supplement them). Imagine an online resource where the course lectures are available not in 50 minute chunks, but in 2–5 minute video segments closely matched to a paragraph of the textbook and a video of an expensive-to-duplicate demonstration, with problem sets right at hand. How much better to review the lecture from the professor's mouth as often as need be, rather than attempt to decipher scrawled and perhaps incomplete or inaccurate notes.

The same tactic can be used at an altogether different level. Infrequently taught ancient and medieval languages (such as ancient Syriac or medieval Occitan) are in danger of disappearing from study. Even where faculty have the skill to teach them, they are often not given the time to do so in their normal teaching load, while many institutions have no qualified faculty for many such languages. If self-paced interactive instruction, with endless drills and exercises, were available online world wide (there is no technical obstacle to doing such a thing today), then that local faculty member could monitor a student's progress at the outset and spend face-to-face time six months or a year later taking the successful student to the next level—a luxury that few have today.

Additionally, resource-based learning is immensely powerful for distance learners of all kinds. The market for higher education among people whose lives do not allow them the regular assignment of time and presence that traditional teaching requires has hardly been touched. I have taught Internet-based seminars on Augustine and on Boethius with hundreds of auditors and paying customers from around the world who got course credit from my university for rigorous work carried on far from Philadelphia. This experiment suggests that there is an appreciable market for an arduous discipline like classics.

The secondary school Latin teachers of America, for example, work often with little contact with each other or with the academy, and they are too few and too scattered to justify classroom-based course work that can reach more than a fraction of them. But in the aggregate, the Latin teachers of America are at least as motivated and probably better qualified to take advanced work than our regular undergraduates. If we can deliver high quality instruction to them reliably via the internet, we do ourselves a favor (more students), we do them a favor (reenergizing and redirecting their teaching), we do our profession a favor (building from the school level up), and whatever benefit the study of the ancient languages confers on society as a whole is measurably increased. In addition, somehow—and perhaps this is the most important point of all—the joy and the wonder of it all, the magic of education at its best, spreads farther and deeper than before.

The same power over distance and the integration of text, image, and sound has many other applications. Not only within the institution but also between institutions it becomes easier than ever to cross boundaries of geography or discipline. We can build true research seminars, where specialists in disparate locations maintain a continuing interactive discussion at a high level (which is now confined to the exchange of written papers and occasional conferences). Conversely, it is possible to imagine a stage when it will be more feasible to build research teams in one location, bringing together on one faculty of classics, say, half a dozen Greek historians. This luxury, which no institution dares allow itself today, would be possible if the students in that institution who wanted to study Greek rhetoric could supplement their instruction with the teaching of scholars at other physical sites. As it is now, we need to replicate in dozens of institutions small, disjointed classics, German literature, or astronomy faculties, with one each of several kinds of specialists to cover our teaching loads, but with undoubted inefficiency in promoting just the high level of scholarly discourse that our institutions exist to create.

No one should try to substitute this kind of technological teaching for the old vision of Mark Hopkins on one end of the log and the student on the other. Technology can be dehumanizing and distancing. But we need to be more honest with ourselves in higher education than we customarily are about the quality of teaching right now. Too much of what transpires in higher education here and abroad is already dehumanizing and distancing. Stringencies of economics and defects of human character already subject our students to huge lectures, novice teaching assistants, itinerant part-time lecturers, and other makeshifts. Where the ideal relationship between teacher and pupil exists, we might be tempted to think of strengthening it but should not try to supplant it. But there is more than enough imperfection in our endeavors to give us the opportunity to apply our new tools to give education a more human face.

And of course, education is not just downloading information. The most valuable part, all agree, is the personal contact that motivates, ignites, and guides. American higher education has long struggled with the right model for facilitating this kind of connection between faculty and students. Woodrow Wilson's preceptorials at Princeton, somewhere between an Oxford tutorial and a German seminar, are an institution that all who know it praise and all are quite sure died some time ago—even, or rather especially, the most skilled contemporary practitioners of that local art. But it offers a model for what the professor of tomorrow should be doing.

For the professor of tomorrow is no longer what he was in the days when the university embodied all studies in a single location. The university was once a microcosm, a miniature world offering the whole of knowledge in a restricted arena. Every discipline represented had its professor who was the supreme local authority on the subject. That supremacy faded long ago as the growth of our great libraries over the last century brought the world to our door and students found more ways to learn about their subject than to sit and listen to the local professor, but the structure of our institutions of higher learning still reflects that origin. The old model may still be powerful and useful, and we should think carefully about how to adapt it to the future, while remembering that new metaphors can be useful as well.

The real roles of the professor in an information-rich world will be not to provide information but to advise, guide, and encourage students wading through the deep waters of the information flood. Professors in this environment will thrive as mentors, tutors, backseat drivers, and coaches. They will use the best skills they have now to nudge, push, and sometimes pull students through the educationally crucial tasks of processing information: analysis, problem-solving, and synthesis of ideas. These are the heart of education and these are the activities on which our time can best be spent. Apart from that, the professor will be a point of contact to a world beyond the campus. The image I like is that of the university as a suite of software, a front end, or what you see onscreen and interact with, to the world as a whole, chosen for its power, speed, functionality, ease of use, even for its user-friendliness. The professor turns into a kind of software icon—click on the professor and let him take you to the world that he knows.

The power of the World Wide Web is a tiny fraction of what networked information will bring. But a well-structured homepage can already be a place where the student finds everything from a course syllabus to an extended reading list to the handouts that used to litter the classroom floor after a lecture to sophisticated research archives. Such a resource can entice each student to go just so far as they can, and perhaps quite a bit farther than either student or teacher thought possible. Boundaries fall and blur, and the capacity of the curious is the only limit. When the professor is available online through email or real time conferencing (and no longer only when the door is open for limited and busy office hours), the electronic interface will sharply improve the personal dimension of education as well.

Somewhere about now, if you are like me, you will begin to feel a little dizzy at the prospect before us. There is no doubt that our future, like every future, will take as well as give. The promised land does not lie around the corner now, and it would be silly for anyone to think so. There are things we cherish about the face-to-face intimacy of our institutions that we will undoubtedly lose. But we regularly sacrifice intimacy to achieve freedom or power, and we've made such choices in one form or another for centuries. Every technology of the word from the invention of writing to the invention of the Internet has given those who use it new range and power and intimacy of one kind, but dissolved a little further the physical bonds of face-to-face community. There is similar loss and gain ahead, no question.

Perhaps the most perplexing question affects reading. Do we still read as closely and carefully as we once did? Will we in the future? Cautious voices are heard, recently that of essayist Sven Birkerts in his *Gutenberg Elegies,* lamenting the loss of a culture that goes with the loss of close reading. As an impassioned reader, never truly at home save in a swamp of half-read books being devoured by turns, I can already feel the nostalgia with which I may yet console my old age, gloomily surveying a generation of students and scholars who infosurf instead of read.

But on this point there is no turning back. While I've spoken against determinism, I do insist that a technology this powerful will not be refused, no more than writing or printing were in their day. The institutions we inhabit will transform themselves or fade. I remember an old story about a farmer who had an axe he liked—liked it so much he'd put two new heads and three new handles on it, and kept "it" in the same place he had always kept it. The joke is supposed to make us laugh at silly farmers, but it plays on human ideas of identity. Continuity gives identity as much as essence does, and if the first form of that axe had rusted and broken and been tossed away, the farmer would be axeless.

The universities of today, however, strongly disresemble those of five hundred years ago and thrive on a transformation that embodies a lived experience of continuity, wherever the journey has led. Where we would be without the universities that have managed their way through the last five hundred years of change we cannot say. The question now is whether we keep adapting them, installing new heads and handles as needed, or try to keep them as they are.

To decide what, as individual practitioners, patrons, and students of the humanities, we will try to make of our parts of the universities, we must first decide what the humanities are for. There are always those who hold, in the face of substantial empirical evidence to the contrary, that the study of the humanities makes people better, more virtuous, more wise. Such study can do that, but it is not the only way to achieve those goals, and it is far from a certain path. Whether the incidence of virtue and wisdom in our academic midst is sufficient to justify the costs is something that can reasonably be debated. To believe in the virtuous efficacy of these disciplines, we must believe in some kind of dislocation and transference that is not at first sight obvious.

Closer to the bone is an approach to the humanities that emphasizes what the study of cultures and their products can do for the quality of our discourse in and about our own culture. There we need believe less in an invisible transference and more in a direct connection. The study of antecedents is inevitably a study of models and cautions, precedents to imitate and pitfalls to avoid. Even among the vicious, the humanities can demonstrably beget attention to detail, patience with argument, respect for disagreement, a concern for the opacity of the self to itself and of other minds to ourselves. In the end, the study of discourse teaches us to recognize language for what it is: the bearer of a culture that is neither to be despised nor idolized, something to be cherished for its excellences, and amended for its faults.

Can we find more flexible and constructive ways to imagine our world? Will the quality of our public discourse improve as our tools of communication broaden and deepen the community sharing that discourse? These are questions urgent well beyond our campuses. If we cannot address them usefully, our claims about the value and dignity of academia will be increasingly dismissed.

The degraded yammering of the culture of the mass media can only instill in us a hope and a caution. The hope would be that the artificial public space of the mass media will dissolve and disappear and take with it the frauds who practice there. The caution is that what may succeed it is not a single public forum but hundreds and even thousands of overlapping forums. How we make a public life and a community out of that is a question rich with possibilities and rife with risks. The only prescription for such a future is to equip it with citizens as resourceful and circumspect as possible in all the arts of discourse, both the old and the new. Perhaps what is wrong is our assumption that a society needs a single coherent central community of discourse in order to function. It certainly can be argued that the shaping of such central communities has damaged the ability of real historic societies to find a place in themselves for people and communities at the margin. But would not a society without a common discourse be a chaotic thing? It does not take a cynic to retort, "as compared to what?" If we succeed in creating a commonality at the level of social infrastructures, we may need less and less of the self-conscious, problem-solving, meddling superstructure that once seemed necessary.

We do not know what will become of our society because we do not know with any certainty what will become of us. Michel Foucault famously scandalized his readers a generation ago in *The Order of Things* by suggesting that what we called "man" was only an imagined thing, an inherited cultural practice, one that he saw dissolving in our times. Was he right? What if we are not who we think we are? What if thinking about who we are—the theorizing about ourselves—has shaped us, but proves to be no longer necessary? If so, one last good task for traditional humanistic scholarship might be a "history of the soul" and the anthropocentric, rational universe it implied.

Late antique thinkers were preoccupied with the analysis of the soul as the thing about "man" that defined and animated him—for all that it was invisible and impervious to ordinary investigation. That person, the being with a soul, began to be rivaled by other versions a good long time ago. But we still lived in the world which answered to a person's needs, even as the scientists began to press human reason to the point of discovering a singularly inhumane universe. If the humanities have held their own and shaped the "Two Cultures" (C. P. Snow's name for the way the sciences and the humanities deal with each other) by their insistence on an old model, perhaps their future lies in a recognition of the artificiality of all humanistic creations. That recognition may also turn into celebration of the best of humanistic creations, and at the same time liberation from the worst of them.

To recognize the artificiality of humanistic creations is not to delegitimize them, but to dissolve their inevitability. The postmodern world we live in is not an elective community, nor is postmodernism a doctrine we are asked to accept or reject at will. Jean-François Lyotard's seminal *The Postmodern Condition* began life as a commissioned report to the governing authorities of the universities of Quebec; now, almost two decades later, it deserves an audience among the university administrators who have shied away from such seemingly abstruse theory. It is what has become of us, for we are all eclectic hypertexters today, infused whether we recognize it or not by a postmodern esthetic and a postmodern ethic. Channel-surfing, multiply costumed for assorted occasions, multi-distracted, mall-shopping, and role-playing, even our most traditionalist leaders are fully postmodern in their manner of traditionalism.

We are left with a choice, between choosing an idol along with Nietzsche or finding a calling. Good and evil are not questions to be solved for us by our studies but questions to be solved by the enactment of our lives.

The humanities are a repository of what has been done for good and ill. If they are only that, they fail. If they also give us tools and energy and the possibility of enacting a more powerful sense of connectedness to all those with whom we share the planet, including those lying well beyond the traditional bounds of the humanities within our culture or beyond the bounds of that culture, then they are worth pursuing.

Hyperlink

HOW DOES TEACHING WORK?

I have a story of an old teacher I cherish that sheds light on teaching in two contexts. The first is in how I came to tell the story. My colleague at the University of Pennsylvania, Al Filreis, was holding office hours one Sunday evening and invited me to sit in; a motley crew of his students in a modern poetry course turned up and we talked about one thing and another. Time passed and I had to leave, but as I was leaving, I thought of this story about Mrs. Shoppach and promised to tell it to the group.

At the time Al was holding those office hours, I was in Seattle, Washington, not Philadelphia; he was in upstate Selinsgrove, Pennsylvania, not Philadelphia; and his students were scattered all over the University of Pennsylvania campus. We "met" in a MOO, an online conferencing space. We'd all rather have met at the New Deck Tavern or some such familiar hangout, but it was eleven o'clock at night, Al and I were far away, and this lively discussion we had was far better than no discussion at all. At any rate, when I logged off, I went away to write my Mrs. Shoppach story, sent it by email, and Al added it to the repository of online materials for his students to read, even the ones who didn't get to office hours that night.

The story itself is instructive about teaching, because Mrs. Shoppach was a good teacher. This is the story as I wrote it for Al Filreis's students.

From: "James O'Donnell"
Subject: Mrs. Shoppach
Date: Mon, 17 Apr 1995
00:08:19–0400 (EDT)

In grade four, Mrs. Shoppach would give us penmanship exercises. She would write a poem on the blackboard in her fine Palmer method hand, and we had to copy it, in pencil. The rule was simple: NO ERASURES. If you erased once and handed it in, you got an F. Very simple.

Now you can see what she was doing with 9-year-olds. Getting them to slow down, calm down, control their stray bodily fluids, get a little discipline. Makes sense.

So we'd sit there copying, then after a few minutes, you'd hear "Rats!" and a kid would crumple his paper and start over. You had to get it 100% erasure-free or you had to come back after school to finish it.

Now this exercise taught us two very useful things: (1) good penmanship, and (2) truly extraordinary erasership. Because, face it, you can't copy "The Village Blacksmith" perfectly every time at that age, and it was a bore. So you studied the situation, bought good fresh soft rubber erasers, learned to write lightly without pushing grooves into the paper, learned to erase carefully around the blue lines on the paper without abrading the surface of the paper—she'd hold the paper up to the light and could see where you had abraded it. So the ideal erasure just lifted the pencil mark off the paper and didn't touch the paper at all.

Most of us got pretty good at it. She only caught me once, and gave me the only F of my grade and high school careers. (Kids crowded around and cheered, but that's another story.) I looked at the paper with the F on it and did not say, "Gosh, I have sinned, I must go and mend my ways." I looked at it closely and said, "Geez, she's good! I gotta get me a better eraser and work on dealing with those crossing strokes right at the blue line." And I did, and she never caught me again (though I suspected her then and suspect her still of sometimes letting really good erasership pass.)

Now on long reflection years after, I realized that the genius of that teaching strategy was that both parts were important and useful. It's important to learn how to calm down and write neatly, but it's also important to learn how to cope pragmatically with unrealistic demands on your time and talents. How good do you actually have to be? Which assignments can you shortchange? How much of the reading do you really need to do? THOSE are real-life, real-world skills, because you're going to be juggling multiple demands forever. And indeed, in that case, the skills of erasership that we learned were exactly congruent with the skills of penmanship—they directly helped us produce written work that was neater, cleaner, more legible, etc.

My point is the one that I was making in the MOO tonight. The content of what we teach is one thing, but the form in which we teach it, the way we manage the classroom and the assignments and the evaluation, all those are equally important parts of education. A chief piece of what you do in college is learn how to juggle several courses and your life without anybody checking up on you. Now in the classroom, the traditional lecture format seems to me absolutely dead; the seminar that leads to people going away and writing (for graduate seminars, often writing them weeks or months later) private little obsessive papers proving how smart they are—there's some use to this, in that you learn things, but the behavior that these practices inculcate is of no use, and in many ways wildly counterproductive, when applied to the real world. My challenge to Alph [Al] is to go on thinking and working on how we adjust our practices as well as our content, to optimize results.

jo'd

Postscript: I met Mrs. Shoppach again at a grade-school reunion in 1995, thirty-three years after I had last seen her, and she confirmed my memory of the writing exercise and still claimed that she never knowingly cut anybody any slack. I'm not so sure about that: I don't think I was that good. But thirty-three years later, she also confirmed my memory of her force of character: she was the rare teacher whose influence runs wide and deep for a lifetime.

Chapter Eight

WHAT BECOMES OF UNIVERSITIES? (FOR PROFESSORS ONLY)

There is something about academic life that drives men (and as near as I can tell, almost exclusively men) to write books about it, sure their readers will be content to hear a story they already know. In recent years, this line of work has mainly been taken up by gadflies delighted to tell us all that is wrong with our institutions, and we have seen a weary procession of profoundly superficial books trudge on and off the best-seller list and in and out of remainder piles: *Closing of the American Mind, Profscam, Tenured Radicals, Killing the Spirit, Illiberal Education, We Scholars* . . . The sum and substance of the indictment is that, though our universities are the envy of the world and represent the sector of the world economy in which American domination is still most nearly unchallenged, we are doing it all wrong, in full betrayal of all we (should) stand for. Ordinary human gormlessness and shortsighted good intentions, the two sources of most misadventure in history, are evidently insufficient to explain the monstrosity of our crimes, and so an ideological failing must be sought. The defenders of academe seek mainly to turn the charge of ideological error back on the attackers. A professor who writes on these topics to a general audience without evident ideological attachment is readily dismissed as a defender of the status quo (and probably given by imputation membership in one ideological camp or the other, or perhaps—depending on the reader—both). The most fruitless and outworn of empty quarrels, that between left and right, is reerected and played out, like some Civil War pageant with all the life gone out of it, over and over again.

In an attempt to avoid the ideological battles, I am indulging the temptation to write of university matters in a way that addresses myself only to my fellow academics. Other readers are welcome to kibbitz, but "you" and "we" in this chapter are confined to the denizens of America's campuses.

The fact is, we're not an industry a reasonable person would invest money in by any conventional criterion.

Consider the facts: Individual institutions are locked into a no-growth policy and our productivity is flat to declining. Tradition dictates that if a state system requires more capacity, it must build more campuses. Indeed, growth on a given campus usually consists of growth in the size of the student body, not an increase in any other facilities beyond the bare minimum, and so is unwelcome. Teaching loads (one measure of productivity) have been reduced over the years, and the characteristic reaction to prosperity is to reduce productivity further. And there is outrage among the faculty ranks and talk of emphasis on "putting fannies in the seats" when deans ask us to teach a few classes that enroll more students. (One anomaly here is that we measure teaching productivity by the industrial credit hour system which assumes that all education is reasonably broken up into equal sized and equal weight units, thus that a seminar on postmodern film theory necessarily counts for as much as organic chemistry boot camp, and a course with 600 students counts for the student as heavily as one with 6.)

To be sure, our other productivity, research productivity, is important to us, and by quantitative measures has doubtless increased. But we don't even know how to be happy about that: there is more talk of an uncontrolled flood of needless publication than there is of abundance and prosperity. The research production side of the university brings definite financial advantages (in grants brought in, for example), but also real unbudgeted or underbudgeted costs. When the university pays a faculty member to do research and write an article, for example, long tradition then allows the faculty member to give the article away, or even to get the university to help *pay* publishers' "page charges," to a commercial publisher who then sells it back to the university in a journal.

Our organizational structure often leads to self-absorption, internal bickering, and feudal kingdoms. The prevailing model for our behavior is individual and entrepreneurial within a socialist environment. We speak highly of the value of teamwork, but the way prestige is distributed in the university discourages horizontal teamwork, fostering occasionally instead the vertical teamwork of lead professor and research team of assistant professors, postdocs, and graduate students. And since academic departments consist (in principle) of independent and unbossable individuals with common responsibilities and goals, little more than moral suasion is available to chairs and deans to encourage discharge of those responsibilities by sometimes reluctant department members. Such teams as we do form in those departments are as sluggish to turn as the Queen Mary. Our deans gnash their teeth and lash out at us as best they can in what they see as the only way to influence our behavior. But we're quite sure that the lashing out is an inappropriate reaction to our behavior. The imbalance of resources as distributed between research and teaching on the one hand and between the sciences and the humanities on the other can be enormous at some institutions. But a department is a department, and when you lump enough of them together, you have a "school," the irreducible feudal component out of which universities are made. There is no longer a center and margins, but multiple centers, presumed equivalent, but of widely varying natures. Against an old idea of the unity of studies, administrators now win wide acclaim if they obey a mandate for diversity and interest, but evince no special concern in the survival or prosperity of any given unit save the profitable.

In addition, we give higher grades than we ever did, feel badly about it, and nod sagely when our colleagues or the public complains. No student ever complains that grades are too high, of course, but no one knows how to make student success into good news. (Discussions of grade inflation are ideal academic discourse, for there is no reasonable possibility of any action arising that would alter the situation we deplore.)

Our students puzzle us. We are ready to teach them the languages that would let them understand our multicultural world or the sciences that would empower them to understand and shape the material world, and yet they divide themselves between the most narrowly pragmatic studies (business, premedicine, and the like) and the most mildly vapid (the uncontrolled smorgasbord of the history major taught in huge lecture courses). We teach art history as our elders did, and fume at students who don't pay attention when we think they should. Our response to these experiences is divided between routinizing despair and intensification of method: if they don't get it, try harder, speak louder, insist forcefully. The results are refreshingly ineffective.

Teaching has something to do with why we are there and especially with society's willingness to pay us but few of us have been trained to teach, and when we have been trained it is by specialists in our own disciplines, not in teaching. We actually care deeply about our teaching, most of us, work hard at it, and are frustrated when we have a collective reputation for being careless and inattentive (of course, in the absence of supervision, many of us are), but our powers of intellection are too limited to conceive a way to improve the quality and responsiveness of what we do. We think it would be vastly undignified for us to invite the psychology department or the education school to consult with us about our practices and offer advice and guidance. And in any event, it is not clear how such consultation could take place, for the classroom is the most private place in the university. Our respect for the sanctity of our colleagues' classroom (save only when perhaps we sit in ceremonial judgment on a junior colleague) would make such consultation difficult. Students do fill out evaluation forms, and these are read assiduously by chairs and deans, believing that the judgment of the unskilled is the only measurement we can make.

Finally, we *do* manage to concentrate our attention (and attract that of the world) on our disputes about curriculum. In so doing, we have put ourselves in the position of being caricatured as left-wingers and, more astonishing yet, put in the position of being accused by the American right wing of seeking to restrict freedom of thought and discourse. It scarcely seems possible that we could have put ourselves in this position save by main force of concerted effort, and yet we show every sign of having done so effortlessly, thoughtlessly, and quite naturally. There is no credible evidence that a required or encouraged curriculum for general education has any effect one way or the other on educational development; accordingly, we have chosen to fight it out to the last ditch over issues of what will be included in such a curriculum. For a long time, literature and philosophy have been the focus of such wrangles, but we have lately found—with the financial support of Lynne Cheney's National Endowment for the Humanities no less—a way to foul the history curriculum with the same tar-brushes.

Meanwhile, we have lost sight of both the students and the world beyond. The organon of studies in the modern university's school of arts and sciences was set in concrete some time ago. Odd bits of inter-disciplinary filigree are added from time to time, but the way it made sense to divide up the world a hundred years ago is in the main the way we divide it up today. That these divisions often mean little to our students or to their experience of the world outside is as nothing to us. The academic left, if that is what it is, persists in teaching what it wants to teach, while the academic and nonacademic right insist that we teach what they want us to teach, and the argument is facilitated by both sides' agreement not to worry about what students might want to study or what their studies might have to do with each other.

But in the end we know, in our heart of hearts, that we have two great advantages, two ultimate pillars for the socialism we despise and cling to, two measures of our worth to society. The first is that we have arrogated to ourselves the right to issue credentials not only for professions and trades that require specific validation of authority, but for virtually all of economically prosperous life. American marketing has had many triumphs in its history, but the creation and selling of "the college degree" is surely among the mightiest such achievements. As long as we control the gateway to the middle class and beyond, we are sure of some survival.

And our second advantage is like our first in that it is also a monopoly. While the American university is a powerhouse of research and a dynamo of instruction, it is also the most delightful youth camp in the world. We should not minimize and we should certainly not ignore the powerful truth that universities are also socially important places where a substantial segment of our youth pursue, for the most part quite unsupervised, an essential stage of sexual and social exploration. To be in residence in a great university is, for our most fortunate 18–22 year olds, a golden halfway house, with most of the benefits of adulthood and most of the protections and privileges of childhood. The collegiate experience was created for the benefit of the scions of wealth, but from 1950 to about 1965, we mass produced that privileged life and made it available to millions of young people each year. Our students may think they should go to college to get an education and to suit themselves for a career, but they also want to go to college because it will be fun. (And when they think of "fun," nothing that a professor does to, with, or for them comes readily to mind.)

The underlying dilemma is that a university is not a corporation but a community whose function is to provide a stable platform for the individual enterpreneurship we call scholarship. But the institution as an institution is losing and will lose its ability to provide the stable platform if it does not respond as a corporation would. In the end it may be a more frankly financial entrepreneurship of units within the institution that enables them to survive. Yet the most familiar, traditional, and central parts of the university have the fewest opportunities for that kind of entrepreneurship. How do we get the business school to care about (and pay for) the department of Egyptology? If we don't succeed in that, do we do without Egyptology?

Before we examine what the university, traditionally a creator and purveyor of relatively scarce high-quality information to various clienteles, must do when it finds itself living in a society with more information than it knows what to do with, we need to take a deep breath and get some perspective.

The earliest complaints of infoglut are downright ancient. Seneca, the Roman philosopher, mocked people in the first century C.E. for owning so many books they never had time to do more than read the labels on the outsides. In the early seventh century C.E., Isidore, bishop of Seville, contemplating the mass of Augustine's surviving writing, uttered the epigram, "The man who says he's read all of you [Augustine] is a liar." Five million words accessed in the animal skin interface common in that time is enough to give anyone a headache.

The same observation can be made of Mark Twain in *Life on the Mississippi* describing how he had to learn the Mississippi River as a trainee pilot, to know every eddy, current, and landmark for some seven hundred miles and more, and to know how they would appear and affect his course going in both directions—effectively doubling what he had to know. While for me the Mississippi River is nothing but scenery and a pretext for bellowing "Old Man River" in a tuneless baritone as I cross over it, to Twain, it was information he needed to have access to, to have learned. Similarly the volumes that contain the five million words of Augustine in our libraries are priceless information to me, but mystic runes of little interest to all save a handful of my colleagues in my university.

My experience as a human being, not to mention my experience as a scholar, tells me that I cannot ever possibly imagine thinking I have enough information. At no time in my life have I ever had enough time to get all the information I need, and I am wearily used to making compromises in pursuit of adequacy. That quality has nothing to do with the abundance of what is out there, and has everything to do with intellectual character and discipline. What is perceived as infoglut is mainly infoguilt—a sense that I should be seeking more. Well, we always should, but we make choices, we filter out noise, we select high value information and we make our best combinations.

This point is worth emphasizing because it highlights what business we academics are in. I've mentioned the railroads of the 1950s before. They thought they were in the railroad business, and they were perilously wrong. If we think we are in the information business, we make the same mistake of confusing a tool with a goal. The real strength of the professor has always been as an organizer, an evaluator, and a processor. If our society reads its infoguilt as infoglut, then our value as information organizers and presenters should on principle go up sharply. Will we be clever enough to see that and make the marketing moves necessary to take the advantage? On our track record, I wouldn't bet the ranch on it.

But what would it entail if we could do that?

It means first embracing a wisdom we all recognize but practice reluctantly. Wisdom means learning to make do with less, or rather, discovering we can do very well with less, and in that self-denial create or find resources to do more than before. This wisdom can be imagined even when it cannot be achieved. In a world that seems increasingly rootless, that wisdom consists of knowing what our important links are and what we can do without.

One discernible and consistent impact of technology on social organizations is the attenuation of social linkages. Inch by inch, the network of face-to-face contacts of the primordial village has been thinned out and dissolved as more and more rarefied threads link us to people farther and farther away. This is a disconcerting process, and one that repeats itself indefinitely. Our universities are themselves products of that process, bringing together people from diverse places and classes and giving them a focus of common interest and activity. In the yin and yang of this attenuation, human beings naturally work at making the new linkages stronger and more nestlike, and so set the stage for the next stage of attenuation.

The metaphor of "roots" is only a a metaphor, but a powerful one that we often use to confuse ourselves. We speak of rootlessness when what we mean are only fewer roots, but perhaps, if we are lucky and resourceful, more tenacious and useful ones. And if we have fewer roots—linking us to an authorizing past in our present society, we assuredly have more links outward. Most people, and certainly most professors, have far more and far more meaningful connections horizontally among our contemporaries than could ever have been possible in earlier generations, and we are only just staggering under the impact of discovering that these connections can span a continent and a world with email in a matter of seconds. The constellation of people we "talk" to will surely change dramatically, and people close to hand, with offices on our hallway, will drift further away as former strangers across an ocean loom larger in life. (Unless of course, we find that we spend more time communicating with the colleague two doors down by email and face-to-face than was ever possible before.)

There is a challenge in here for our universities. Once it was clear and unavoidable that you needed to come together in specific face-to-face communities in order to pursue or transmit knowledge. The physical fact of the library, the great technological achievement of nineteenth- and twentieth-century education, was for many disciplines a compelling focus. The logistics of bringing students together and managing their affairs in bulk was the further, and typically nineteenth-century, rationale for building our university communities, even when they ballooned to include thirty thousand students. But what happens to the economic basis we have formed in these circumstances when students begin not to need so much of our presence or when they can spare us less of theirs? If we continue to provide only for those who come to us, and only for those who submit to our four-year credentialing disciplines, we may reasonably expect that other paths will be found to do some of our traditional business. We must use our imagination to remake ourselves as a larger, virtual presence in society.

As we look to the future, we must be very clear about how we do our business now. On the surface, the university looks like a marketplace where customers select and pay for what they want. But many of our customers experience the university as more restricted. Students are not allowed simply to shop wherever and whenever they will: they apply for consideration, and only if they have the qualifications to match one of the limited number of spaces available are they allowed inside. There may well be comparison shopping, and students speak of "taking the best offer," thinking of scholarships and the like as compensation rather than as discounted prices. Once inside, they submit themselves to a schedule and to tasks set by others, with evaluation coming from authority figures who have the right to dismiss them for poor performance.

In some places, higher education is already using a more frankly consumerist model. There are community colleges who deliver classes to their students, or at least to their neighborhood, and who respond strictly to demand and market pressures. Distance learning programs and continuing education programs exist to find and respond to markets that have been neglected. Struggling small colleges have discovered that though they wish to seem selective and traditional, a fair amount of often desperate marketing is necessary to stay in business. In the Ivy League, a scholarship is still an award; for a small poor Catholic college, it's frankly a discount and a marketing strategy. This consumer-orientation will increasingly impress itself on a larger and larger sector of higher education. We must realize that our rhetoric of marketplace has not matched our practice, and the change in practice to draw closer to our rhetoric will prove unsettling.

Most threatened in this change is the traditional role of the teacher/professor. The envied and controverted status of tenure is only a reflection of our old expectation that teaching is not just a job to be done for a time, but a career and indeed a vocation. Professors are almost the last members of society who think they can afford to train themselves when young and then pursue a single linear career for the whole of their working lives. Almost everywhere else in middle-class American society, the security of such linear careers is unraveling daily in a welter of downsizings and corporate takeovers. If professors seem immune for the moment, they should not take too much comfort in that appearance. Outside traditional educational institutions, of course, teaching is a job like any other. Corporate education programs have no tenure and little continuity beyond response to the immediate needs of the company. The differences and the similarities in what they do highlights again the idiosyncratic homogenization of our higher education institutions.

One question we must ask is if it is clear that the imparting of facts and the inculcation of skills needs the expectation, the prestige, and the protection of tenure. Tenure is surely at risk, can we see that and still find ways to offer the protection at least when it is genuinely of value? A businessman would observe that a fringe benefit can be offered to employees as long as no serious competitor does without it. Comfortable American labor union contracts began to look extravagant when and only because other competitors emerged who did without them and could beat the employer's time. Can this happen to universities? Are we vulnerable, as the rest of the American economy has been, to unanticipated competition?

Some forms of competition can already be seen emerging. The "training sector" of our economy has expanded beyond business settings to provide services for traditional students outside traditional educational institutions: MCAT-preparation courses, for example, do part of the traditional institutions' business, and it is not hard to imagine a setting in which a voucher program empowers parents to decide just where to get the educational product their children need. Many students already take a minimal curriculum with a maximum of easy courses—at Ivy League institutions there is a tradition of marginal students taking the most difficult required premedicine courses in summer school at less demanding institutions—and then spending money on an MCAT-preparation course to make sure they get good MCAT scores. A good education is a beautiful thing, but a good MCAT score is worth paying for. As long as *we* in academe choose to emphasize our credentialing role, then we are vulnerable just insofar as more direct and economical paths to those credentials can be devised.

The real competition is likely to be subtle, and it comes from ourselves. It is there already in the burgeoning community college market, for example, that offers people two years rather than four, convenience of hours and location, low tuition, and a discernible path to specific skills and credentials. It is there already in the increasing number of students who do not follow the four-year lockstep, but go to college with a weather eye on the job market, seeking employment well before they have the degree, and postponing the degree when good employment comes along. If students begin to sense widely that the value of a degree is so limited that it would be silly to pass up employment opportunities along the way, then the old myth that "To get a good job you need a good education" will begin to crumble and with it a part of our power base in society.

In theory we can imagine the final competition arising when we ourselves begin hiring the uncredentialed to do our teaching for us. If we were to go so far, this argument would say, as to hire skill and utility without regard to certification, the ball game would be over.

But perhaps it is already over. No development in higher education in the last twenty years is so likely to have long range effect on our institutional structures as the move to do more and more teaching not only with teaching assistants (the invisibly uncredentialed whom we take for granted) but with impermanent, nontenured, non-tenure-track faculty. Pieces of the curriculum are handed over, one at a time, to the marginal faculty. And it is the most marketable part of what we have to sell that we give to them.

We have decredentialed these people by denying them the opportunity to use their credentials in a traditional way and they compete with our students for careers that are increasingly denied them. Rather than construct institutions that responsibly offer the next generation a place at the table, we offer a subordinate role without hope of permanence or authority—a life for gypsy scholars. When we worry about this, we do so in a characteristically sentimental and self-regarding way, wondering whether we are being sufficiently virtuous when indeed we should wonder whether we are being sufficiently selfish. This underclass of teachers is the gravest threat to our status as mandarins.

But there is one further form of competition that we should anticipate: when we turn on each other. It is not impossible to imagine that the humanities and social sciences will continue to wrangle with each other and with ideologists of all stripes and that the scientists who have looked on these struggles with disdain will discover that their status and access to financial resources are being dragged down in the general decline of institutional reputations. If that happens, the warfare of mutual contempt among disciplines that often simmers below the surface in institutional debates could become frank and open. The university of the postwar world will suddenly look like a Yugoslavian mélange of incompatible and mutually suspicious tribes ready to turn on each other in an instant. The ethnic cleansing that a university could imagine will take few if any lives, but could have an equally drastic effect on the unity and harmony of the institutions as a whole. That late antique dream of the unity of knowledge has been long-lived, but may not be immortal.

What principles can help us think our way past these unpleasant scenarios? Both our two residual strengths, monopoly control of credentials and unique supplier of youth camp for the near-adult, are at risk and at the same time offer opportunities.

Entrepreneurs don't need credentials, technical specialists do. Which is the model for our time? For the last fifty years, a technical model has given us a lock on the credentials market for a huge sector of society. But the buzz today is entrepreneurship, and those students of whom I spoke who drift out of our rooms to pursue jobs are using a different model: seizing opportunity and regarding credentials skeptically.

We could do two things as we react: (1) multiply opportunities for study, and (2) repackage, streamline, and market our credentials-giving. To do either of these we must separate teaching from evaluation, the nurturing pedagogue from the hanging judge. The American system that combines these two roles is so deeply rooted that it is not practical simply to suggest that they be given a divorce, but what we can do is find ways to offer credentials in specific subjects separate from our teaching of them, and to offer those through and with the prestige and authority of the university. We may preserve the high nobility of the Bachelor's Degree (if high nobility is what characterizes it) as we like, but there is nothing to forbid us and everything to encourage us to think about increasing the number of special certificates of competence, expertise, and training that we offer. In this way we could encourage the part-time student, the self-taught student, and the commercially-taught student, to continue to rely on us for what they will see as an essential part of their education: proof of having met certain standards.

But certificates and credentials are not education, you will say? Quite true: but if that is a criticism of what I just suggested, it is a fortiori a criticism of the system we now administer, where we offer education and students come to us for the certificates. If we can think resourcefully about ways to begin to separate the two processes, we can do a better and more objective job of evaluation, and then turn to doing a better job of the nurturing and expansive teaching that we like to think we are really good at.

Whether that job is best done in the traditional youth camp university setting is a good question. Our old centers of higher learning are so encrusted with habit and self-satisfaction (and self-satisfaction's evil twin, the kind of self-loathing that knows not how to change itself), that I judge it more likely that change will come at the margins.

Some of these margins are quite close at hand in the distance learning and continuing education branches of existing universities. Faculty who work in these fields often find immensely rewarding teaching to do and the opportunity to experiment, but they typically do not have either the security or the encrustation of the "day school." They are certainly the most market-driven sector of established higher education: there is no captive audience in night school.

Far from the big university campus, another opportunity is already being seized by small private colleges. The United States is remarkable for the huge number of these private colleges that dot its landscape, often in remote and out-of-the-way places. Often they incarnate one or another distinct community of values, religious or otherwise, and represent at the same time a fruitful interplay between the province in which they are situated and the wider world beyond. If higher education's task is to create a space where the student can move beyond home and province into that wider world, these colleges have played a distinctive role. The most prestigious are indeed universally respected.

But in the last generation, since the constriction of resources for higher education began in the early 1970s, the small colleges have found themselves clinging to their traditional niche by their toes. They are small, undercapitalized, overtenured, inadequately flexible, and incapable of competing in range of offerings with their larger rivals. They capitalize with difficulty on their enviable reputation.

The larger institutions know full well the student who wants the options (academic and social) of a huge campus and the intimacy of the small college. A few of our best research universities can almost provide that combination, but only a few. For the rest, students come in the main to the larger places, harboring a kind of longing for what the smaller can provide.

In that longing is the opportunity for those small colleges to emphasize their tradition of hands-on, student-centered education and make it the consumer orientation that will assure them a future. If even the richest and largest university can no longer be a microcosm and offer all things to all students, and if the challenge of teaching is not to concentrate resources but to put students in touch with the resources of the world, then the advantage of the large institution over the small begins to evaporate. The colleges will likely have a harder time than the large anonymous universities in tearing themselves away from the four-year lockstep for their traditional students, but internship, co-op, and other programs designed to send the student away from campus, perhaps indefinitely, while retaining the link to their professors will work for them.

The model of a "lifetime warranty" or a "service contract" from your institution of higher education is tantalizing and some large institutions now fiddle with it. How to adapt it on a smaller scale to give students an intellectual frame of reference for the future as well as for the present is a challenge that can bring not only better education, but also stronger alumni loyalty. While the best of the private institutions already have alumni loyalty that shapes a lifelong funding base, it is a strength they must not let slip away.

To use our new technological tools to change education, we must know what it is we are trying to do—what the purpose of education is. The following reflections arise from my own experience in the confrontation of media and methods and are meant as suggestions, not prescriptions, the fruit of personal experience being shared (on the "for professors only" terms of this chapter) with colleagues for rumination and revision.

First, the traditional classroom is among other things a place for rehearsing behaviors of use in later life. This is masked by some of the antiquarian practices of our institutions (lecture-hearing and test-taking), but it becomes clearer when you look at the laboratory, the seminar, and the discussion section. In all the places where education is most personal, it creates a space in which students begin to behave as adults, to behave as they will when they have real responsibilities. The focus of such work is to draw the student out of passivity into activity. The best classroom is one in which the student begins to think, speak, write, and act in new ways made possible by that classroom.

That classroom is a potentially frightening place because much of our traditional pedagogy depends on the managed infliction of humiliation. When we call on students to perform, we implicitly threaten them with the risk of embarrassment if they do not perform well. We cannot and should not spare our students this risk entirely, for that would reduce the value of the exercise, the rehearsal to face life's more threatening situations. But much of what we do is now hopelessly old-fashioned, offering rehearsal for unreal worlds, and misreading the measure of embarrassment our students can absorb.

Everything thus tells us to emphasize collaboration, interaction, and student activity. The seminar was an advance in its day and still encourages students to model behavior they will revive in conference rooms and offices all their lives: but a seminar that ends with a dozen disconnected papers being written is scarcely a real-world preparation. The seminar can be instead the common workshop or the task force producing the common product—this is indeed the narrow pedagogical genius of the "case study" method.*

Here is where electronic media can help innovation. The better communication, the personalization, and the multiplication of spaces, real and virtual, in which to invite the student to perform can all make this model better. The student who now is unable to perform adequately in the face of perceived threat of embarrassment in class is the one who can be given a place to rehearse out of sight of classmates and teacher, and then be invited back into a common space when ready. The common space can become increasingly dramatized and in various constructive ways competitive, while the private work is all done offstage. To link the work that is done offstage to what goes on in the classroom is the true nexus of teaching and the best place to concentrate attention when teaching goes astray. What is it that we expect a student to do when out of our sight? What incentives do we give the student that will realistically lead to that result? And what connection does that have with what we ask of them when they come before us again?

Second, the implicit myth of our liberal arts tradition is that we prepare students for a wiser, more cultured life beyond the university. In my day that took the form of older students at college advising their juniors to take an art history course. Sitting in the dark watching beautiful things on slides was pleasant, and it gave you cocktail party con-

*The commonest objection to this suggestion is how to grade individuals for a common project. The correct response to that question is to urge again the disjunction of evaluation from the classroom. A well-structured classroom is one in which students will get feedback from both colleagues and teacher in an ongoing dynamic way. If at the end of a term you need an examination to know how a student is doing, my view is that you haven't been teaching well at all.

versation to last a lifetime, though I doubt that idea carries much weight anymore. If we accept the model suggested above that we are rehearsing behavior with our students in the courses we teach, are we truly giving them a liberal arts education? Or are we rehearsing academic analysis at a distance? or the solitary pleasures of the study? As valuable as I think those virtues are, their selling power is not universal.

Instead, we could imagine liberal arts courses that took as their first business the enacted rehearsal of a culturally enriched life. Some material must be gotten at second-hand, of course: no campus has Leonardos enough to share with all its students. But the art galleries, the musical performances, the literary readings and poetry slams for which our campuses are justly well-regarded are all too often crucially separated from the formal academic exercise of the institution as inadequately disciplined. Students are encouraged to attend concerts, but *required* to show up for lectures and quizzes. Even if what is performed or enacted is of lower quality than Leonardo or the Philadelphia Orchestra, we can make the common experience of art, music, and literature itself the business of teaching. (I do not mean rows of classroom-bound students listening to a stereo dutifully, but real rooms with live musicians and arguments afterwards.)

That rehearsal can point at least in principle to real activities that an actual academic institution might foster for its former students. The boom in "alumni colleges" in recent years began as a moneymaker and a tourist gimmick. There is still too much "college" about them, a going back to school, as though school is a place to go to get away from life. But when I took medieval art in college, for example, it was far beneath the dignity of anyone associated with my alma mater to tell me how I might later go to see some of the works of art I studied. For the most part, provenances and present whereabouts were indicated only incidentally. Is it too much to ask us to provide students with concrete agendas and even the prospect of coming back into contact with their teachers to pursue issues raised once they have experienced real works of art or music or literature? Once again, the electronic media at our disposal today promise to link communities in space and in time; we could use it consciously to bind students together in a community of experience and enjoyment.

The canniest among you will recognize that my suggestions for reform simmply enshrine much of what is already velleity in our institutions. We struggle to take back our campuses from the fraternities and the parties, and we have tiny programs to expand academic activities in dormitories. But the fatal line separating "school" from "extracurricular activities" needs wholesale destruction, and the line separating "school" from "life" no less. Simply moving the line incrementally only confirms the initial error.

One way to destroy this line is to be frank about youth camp. Let us make even dormitory life and social life a part of the educational package, expressly addressing questions of how our students live and challenging them to take themselves seriously as adults-in-training. There is, after all, no reason why a dormitory must be squalid, fetid, and strewn with litter. It is the permission of the adults and the mistaking of self-indulgence for liberty on the part of students that creates it. You cannot send in dormitory monitors to change such an environment by main force of fiat. But I know no mainstream institution in which faculty and students together address the nature and purpose of campus life and seek in a friendly spirit of rivalry and collaboration to make every aspect of that life as rich and humane as possible. To find the language and the diplomacy to carry that off would be difficult, but the results would be electrifying.

The future belongs to the institution that both recognizes and creates a market for a new boundaryless, fearless pedagogy, clearly distinct from the process of evaluating and giving credentials. In doing so, we begin to see how we can even draw upon our most embarrassing established strength to make of our campuses a richer environment for learning on all sides. Once again, the oldest and most established of institutions will move the slowest.

It would be preposterous to suggest that what I have said here is a model for some widespread change. I am no fool and have no expectation, hope, or fear that this might be that. But I also believe that change begins retail, not wholesale, and that it works not by fighting evils but by creating excellence, first on a small scale, then larger. The principles and the directions I propose are not per se irrational. They are at most innovative and difficult to achieve in a traditionally entrenched institution. The correct questions to ask are these: Can even a few of us undertake experiments in this spirit, to see what works, to make mistakes, to learn from those mistakes, and to offer encouragement to those who might imitate us or learn from us? If not, why not? (Two hours: please write on both sides of the paper; while you are writing, notice one thing: the electronic networks and tools now coming into our hands are essential and implicit for all that I have said, but I have scarcely mentioned the tools in this chapter, because the conversation must be centered on goals and the real motivations of faculty and students to pursue those goals with the best tools that come to hand.)

Chapter Nine

CASSIODORUS: OR, THE LIFE OF THE MIND IN CYBERSPACE

In 1962, *National Geographic* commissioned Kenneth Setton, then professor of medieval history at the University of Pennsylvania and later for many years distinguished member of the faculty of the Institute for Advanced Study in Princeton, to write an article on "New Views of Medieval Europe" sketching ways in which scholarship had revised traditional views. When I came upon that issue of the magazine as a 12 year-old living in a border city in the American southwest (whose links to that medieval Europe ran more directly south through Mexico than northeast through the United States), it was not so much the texts as the pictures that grasped my attention.

National Geographic had commissioned an artist in the realist-illustrator tradition to imagine key moments in medieval history. Years later I found and bought a copy of this issue in a used bookstore to confirm my memories. The images were lurid and gripping. Women and children huddle in the sight of flames as Alaric and the Visigoths sack Rome in 410; Attila and his Huns banquet in hideous luxury; Charlemagne gawks at the elephant sent him as a present by the sultan Harun-al-Raschid; and the emperor Justinian sits slumped in his chair, despairing of his reign, while his fierce and haughty wife Theodora chides him to make a stand against rioters threatening his throne.

In the years since, I've come to realize exactly what ancient evidence supports each of these pictures, and chuckled on standing in the church of San Vitale at Ravenna, admiring the huge and beautifully preserved mosaics in that church, to remember that the *National Geographic* artist had borrowed his portrayal of Justinian and Theodora, down to details of their dress, from the hieratic mosaics in which they were honored in a city they never saw. The moments chosen for illustration all play key parts in the structuring of the medieval part of the familiar western story. Savage invasions, self-definition of west against east, civilization against barbarism, these are key issues in that story.

And then there was a picture of Cassiodorus. An elderly monk with a beard and a slight stoop shows visitors through stone corridors into a monastic chamber where other monks are at work copying manuscripts. The image was, like the others, familiar in idiom—too familiar. The monks, I recognize after the fact, are too medieval, not late antique. The habituation to the cloister's habit of making manuscripts is too conventionally appropriate to the ninth or eleventh century, not the sixth.

My experience in youth was to feel (wrongly, as events would turn out) that the American southwest was simply a million miles from anywhere, that civilization lay entirely on the east coast. I had an infatuation with Columbia University while in high school, reading books by famous and less famous worthies of that institution mainly from earlier decades: Thomas Merton, Mortimer Adler, Moses Hadas, Irwin Edman, Mark Van Doren, Gilbert Highet. When the time came I was intimidated by the urban grit and danger and chose a more idyllic setting in which to easternize myself. But New York City remained, to this childhood Yankee fan, the center of the world, beyond all question. To grow up in the sunbelt was not necessarily to see a future there. As an adult meeting other refugees from the American west who have become classics professors I have had the feeling that some of us went the way we did because the frontier was too far from an imagined center, and the center itself, western civilization, had an exotic and romantic air about it, an air that it could not have had for the Greek- and Latin-cramming preppies of the old private schools on the east coast.

Almost a decade after I first read that *National Geographic* issue I came back to Cassiodorus. I had by then left the southwest to live in faux-medieval stone buildings on an ivy-covered campus, and found myself rehearsing the role of classicist.

A combination of Catholic upbringing and some happy accidents encountering inspirational teaching in college made me an odd kind of classicist, to be sure. To embrace the classical past was not enough. I had also to work the Christian middle ages into my self-definition, and my choice lay in the earliest middle ages. For all that my politics were conventionally liberal, these were extraordinarily conservative choices, fostered by a university still to a large extent serenely unruffled—in its intellectual momentum—by anything that had happened since about 1950. (Northrop Frye was a perilously advanced literary theoretician to us, this five years and more after the famous Johns Hopkins conference that introduced Jacques Derrida to a breathless American critical community.) My particular preoccupation was Augustine, whom I gladly accepted as a founder of traditions that I received quite transparently.

But Cassiodorus won my loyalty for the short term, as it seemed.

He was the perfect thesis topic: ancient enough for a classicist to tolerate mention of him, medieval enough for medievalists to take interest, and when I came to write my doctoral dissertation, prudence led me to think of how to package myself to please prospective employers in classics departments, history departments, or religion departments (in the early days of the post-60s collapse of the academic job market, where packaging had suddenly become important) . His appeal was, moreover, as a transitional character: preserving the classics for the middle ages, writing history and theology himself.

Cassiodorus turned out to do what I wanted him to do, which is to say, provide me a dissertation topic, get me a job, and get printed in hard covers to help me keep that job. The best review the book received, negative on numerous points but still far and away the most learned and interesting and helpful, was called "Cassiodorus Deflated." I've gone back and reread that review gratefully on numerous occasions, but always pausing to wonder a bit at the title and the concomitant tone of mild disappointment. The author of the review herself forgot what she had read and a decade later in a textbook footnoted my book in support of a more optimistic version of Cassiodorus that my book had been at pains to debunk. Other reviewers were equally disappointed, even when they praised it. One of the most laudatory reviewers years later described it as a "depressing" book when he introduced me to a lecture audience. Clearly the traditional Cassiodorus fills a need in our mental furniture of his time of transition, a need that persists beyond all reasonable refutation and that evinces the power of the underlying implicit story about "our" past.

At any rate, I certainly had at that time some vaguely conscious idea of the congruence of his situation to that of the Christian scholar today, and said as much in the preface to my book. It was a conservative and traditional kind of book, with a few cautious hypotheses and much mustering of erudition, or at least of its appearance. On rereading it at twenty years' distance from the writing, what I notice mostly is that facility and clarity of style went a long way to give it what merit it has, but it was scarcely path-breaking.

And so, after having written my book, I went away from Cassiodorus.

Time passed. I knew the moment I heard a word processor described (by the novelist William Gass on one of Dick Cavett's talk shows) that I wanted one, and I bought a Kaypro II as soon as they came out. The double-density floppy disk that held 191K of data was just large enough to begin to do serious work. My old typing skills (the only bit of physical dexterity to which I can claim credit) were more valuable than ever, and soon I was filling disks left and right. A few more years and the need to access an online database got me a modem, and I started doing email. A year or so later, a colleague had an idea for publishing a new book review journal, and I had the idea of putting it out on the Net, and *Bryn Mawr Classical Review* was born. In the birthing, we clogged and crashed machines with huge files before we noticed that the "issue" of an electronic journal needn't replicate the paper journal's habit of clumping things together for ease of shipment—and so, by accident and stupidity we learned what some of the defining characteristics of cyberspace would be: for example, a synchronous consumption of small units of text.

Curious and obsessive, eager to use new technologies to expand the reach of communication, enhance teaching, and find better ways to publish, I found myself trying more and more things, some pedagogical, some scholarly. People invited me to speak and write, and I was glad to do so, always wondering why there weren't more humanities scholars like myself.

And somewhere along the line I began to see an irony. I had become like Cassiodorus. Not because I was Christian or scholarly, but because I was more or less consciously lending a hand to the task of creating for the people and ideas I valued a usable space in the new technological environment in which we could make ourselves a functioning community. The chapters of this book all address that challenge in one way or another, but it is worth saying more explicitly what that entails and what the prospects are like.

In these pages I have tried to use and consider the past in a way truly responsive to our present. I am an old-fashioned, text-consuming, text-producing gatekeeper of our culture. I write with fountain pens, find when I move that boxes containing books outnumber boxes containing computers by approximately 50 to 1, and find the most severe economic pressure from cyberspace exercised by online bookstores that let me go from my sofa and the impulse to buy to a completed purchase (delivery in two business days) within three minutes. At the same time, I have been given the chance to live on the edge of exciting cultural upheavals. My greatest confidence is that I have a responsibility to use the privileges I have been given to contribute to the wise navigation of those upheavals, even if they leave me and my kind in the end unemployed and unemployable.

And so it is an odd way in which I have come to *be* Cassiodorus. He spent his adult life as public servant and cultured aristocrat. He began in a stable if newly-erected world of Gothic rule in Italy and spent thirty years on and off in the service of his kings. His was a clear and well-defined role, and he had every reason to hope for a gradual withdrawal into distinguished and affluent retirement. But, you'll remember, he had his world shot out from under him. Cassiodorus himself was first torn away from his homeland to live in a privileged cocoon, half-guest, half-hostage, in imperial Constantinople, then wound up back on his ancestral estate. In one way, he got his retirement retreat, but the world around had changed so much that even his estates were a different place. That he chose to make of them a monastery, to build a library, and to copy books was only one of the possible responses to necessary adjustment.

Cassiodorus chose a course that succeeded in placing some new wine in some old bottles. He used the instruments, the habits, and the cultural expectations of the old Roman culture in which he had been brought up to do new things, create a new kind of library. He is not a savior of western civilization, nor should any of us expect to be. He was rather a single, responsible individual helping shape to the limits of his ability the institutions and the cultural tools that his world needed. That he accepted and thrived on the disruption of his life and expectations, and that he succeeded in using his past and his expectations so resourcefully to help him shape a future, are lessons we can all take away with us.

BIBLIOGRAPHIC NOTES

INDEX

BIBLIOGRAPHIC NOTES

These notes both acknowledge my indebtedness for ideas and information and at the same time suggest books and articles that illustrate and advance the discussion of this book in various ways. In a world awash with speculation about the cyberfuture, one book is already a classic and deserves wide and continuing readership and special preliminary mention: Richard Lanham's *The Electronic Word* (Chicago, 1993).

Introduction: The Scholar in His Study

For the portrait of Jerome, see P. H. Jolly, "Antonello da Messina's 'St. Jerome in His Study:' A disguised portrait?" *Burlington Magazine* 124 (1982) 27–29. For the broader story of Jerome's anachronistic long-delayed idealization (usually in a cardinal's robes), see E. F. Rice, *Saint Jerome in the Renaissance* (Baltimore, 1985).

My readings of Machiavelli and More owe much to some of the most exciting recent work on the Renaissance, including Stephanie Jed's *Chaste Thinking* (Indianapolis, 1989), esp. on Machiavelli creating the private space of the modern writer, and Stephen Greenblatt's famous essay on More in his *Renaissance Self-Fashioning* (Chicago, 1980).

1. *Phaedrus: Hearing Socrates, Reading Plato*

Writing is approximately 2500 years older than any of the figures I discuss here, but the abundant written heritage that comes to us on clay tablets in cuneiform script and the comparably rich treasury of hieroglyphic inscriptions form no part of any modern literary or philosophical heritage. To start the history of "writing" from the beginning, sample, e.g., Jean Bottéro, *Mesopotamia: Writing, Reasoning, and the Gods* (Chicago, 1992).

Derrida's reading of the *Phaedrus* passage on writing appears in the essay "Plato's Pharmacy" in his *Dissemination* (Chicago, 1981). Eric Havelock's most accessible works are his *Preface to Plato* (Cambridge, Mass., 1963) and *The Muse Learns to Write* (New Haven, 1986). Rosalind Thomas's *Oral Tradition and Written Record in Classical Athens* (Cambridge, 1989) gives a meticulous portrait of the many interplays between speech and writing in that culture.

The facts of Greek sexual ideology and practice are repeatedly canvassed to throw light on our own mores. A recent legal case in Colorado has thrown up two separate vehement polemics in opposition to one another, one by the legal theorist John Finnis, " 'Shameless Acts' in Colorado: Abuse of Scholarship in Constitutional Cases," *Academic Questions* 7:4 (1994) 10–41; and another by the classicist Martha Nussbaum, "Platonic Love and Colorado Law," *Virginia Law Quarterly* 80 (1994) 1515ff.; on the controversy, see D. Mendelsohn, "The Stand," *Lingua Franca* 6.6 (1996) 34ff. The inability of our academic discourse to offer agreed clarification on such an issue is an important limitation that must be recognized; scholars are not "experts" to whom to defer.

2. *From the Alexandrian Library to the Virtual Library and Beyond*

One cannot speak of virtual libraries without thinking with admiration of real libraries and librarians. One wrongheaded expectation implicit in the contemporary dream is that its realization will obviate the need for buildings, but the virtual library appeals even to those who are building a very large building indeed: G. Grunberg and A. Giffard, Head Librarian and Head of Informational Services for the new Bibliothèque de France, use the phrase as one of many approximations of the future they sketch in their "New Orders of Knowledge, New Technologies of Reading," in *Representations* 42 (1993) 89. That whole issue of *Representations* is devoted to "The Future Library" and offers a fascinating range of theoretical and practical discussion, centered on but not limited to the Bibliothèque de France (successor to the Bibliothèque Nationale of which I speak in this chapter).

Vannevar Bush's original article ("As We May Think," *Atlantic Monthly,* July 1945) is now on the World Wide Web at *http://www.isg.sfu.ca/~duchier/misc/vbush.*

One should be careful not to press the use of the word "library" too hard when thinking of ancient collections; it is used too readily to apply to entities varying wildly in size and nature, and the common trait of being created and used by people who had similar, but entirely unrealistic, ideas about what they were about should not obscure the fact that the "book" was a different thing to ancients from what it is today: not a text for silent digestion, but a tool to help you repeat the words aloud; not a source of "information," but chiefly a repository for wisdom coded as poetry and narrative—an ancient Dewey Decimal System would not have left "fiction" out of play, but would have placed it at the center of the collection.

The story of the Alexandrian library is accessibly if tendentiously told in L. Canfora, *The Vanished Library: A Wonder of the Ancient World* (Berkeley, 1989); a more sober perspective is that of D. Delia, "From Romance to Rhetoric: The Alexandrian Library in Classical and Islamic Traditions," *American Historical Review* 97 (1992) 1449–1467. The so-called "Letter of Aristeas" gives numbers for library holdings (200,000 in hand, with a goal of 500,000) that are of little value, but they give a sense of how close to totality a dreamer might think the library came.

The book practices of the age of the codex may be explored in L. D. Reynolds and N. G. Wilson, *Scribes & Scholars: A Guide to the Transmission of Greek and Latin Literature* (Oxford, 1991); R. McKitterick, *The Carolingians and the Written Word* (Cambridge, 1989); M. T. Clanchy, *From Memory to Written Record* (Oxford, 1993). In the midst of contemporary debates about the "canon" of authoritative text, it is well to remember that they are not new, rather their history is coterminous with the history of the codex. From the early sixth century C.E., for example, we have a list called the pseudo-Gelasian *Decretum de libris recipiendis vel non recipiendis* (Decretal on books to be accepted and books to be rejected), which outlines which books of Christian literature must be read, which may be read, and which should not be read—the papal "Index of Prohibited Books" lay at the end of the path that text opened. Cassiodorus will return several times in this study. For his life, see J. J. O'Donnell, *Cassiodorus* (Berkeley, 1979), available at: *http://ccat.sas.upenn.edu/jod/cassiodorus.html.*

For the coming of print, see L. Febvre and H-J. Martin, *The Coming of the Book* (Atlantic Highlands, N.J., 1976), E. Eisenstein, *The Printing Press as an Agent of Change* (Cambridge, 1979), and (for specialists, but with interesting reference to contemporary technological change) M. Giesecke, *Der Buchdruck in der frühen Neuzeit* (Frankfurt, 1991).

For a first glimpse of how dialogic discourse may supplant the monologue, see in print form A. S. Okerson and J. J. O'Donnell, eds., *Scholarly Journals at the Crossroads: A Subversive Proposal for Electronic Publishing* (Washington, 1995): the book comprises an extended and highly serious email dialogue involving participants in multiple time zones and extending over many months.

3. From the Codex Page to the Homepage

For the transition from roll to codex, see C. H. Roberts and T. C. Skeat, *The Birth of the Codex* (London, 1983). It is worth emphasizing that a manuscript on animal skins was always an expensive artifact. The earliest surviving manuscript of the whole Bible in Latin, the Codex Amiatinus now in Florence, whose pages are roughly the size of our photocopier paper, cost over 500 calves their hides and weighed in its original form about 90 pounds. The portrait of Ezra/Cassiodorus that comes from the Amiatinus has often been reproduced; one version is available at: *http://ccat.sas.upenn.edu/jod/Picts/Ezra2.gif.*

Some form of the overtly narrative codex book will survive, if only for airplane reading; but even there it is to be noted that experimental literature is already playing with hypertext forms (see, for example, M. Pavic, *Dictionary of the Khazars: A Lexicon Novel in 100,000 Words* [New York, 1988], distributed in marginally differing "male" and "female" editions, and with the alphabetical arrangement of entries differing sharply between the English translation and the original Serbo-Croatian). It is only a matter of time before someone publishes a murder mystery that must be consumed as a self-constructing hypertext of labyrinthine proportions. See for an earlier example, Julio Cortázar, *Hopscotch* (New York, 1967), a novel that offers multiple paths through different sequences of the same chapters long before there was practical thought of hypertext. On the intersection between literary and technical velleities, see G. Nunberg, ed., *The Future of the Book* (Berkeley, 1996).

The current state of experiments with new media of publication can be quickly assessed from looking at NewJour, the online archive of new electronic journal publications: *http://gort.ucsd.edu/newjour.*

4. *The Persistence of the Old and the Pragmatics of the New*

The story of the Urbino Bible has a contemporary irony that should not be missed. In early 1993, a small and select collection of precious books from the Vatican Library traveled to the United States for an exhibit at the Library of Congress. I was among the few fortunate thousands who saw the books while they were there, and vividly recall the display of the first page of the book of the Apocalypse in the Urbino Bible just as you came into the exhibit hall. But that is now for me only a memory, and the book itself is many miles away and access is restricted. It is accordingly of great value that the Library of Congress undertook to make digitized images of the book pages on display in their exhibit and to make them available on the Internet. That very page of the Urbino Bible may now be consulted from any suitably connected place on any of the seven continents: *http://sunsite.unc.edu/expo/vatican.exhibit/Vatican.exhibit.html.* On a properly equipped machine, the image can be enlarged, cropped, enlarged again, with remarkable accuracy of detail: I have seen it enlarged on a projection screen to several times its original lifesize with brilliant representation of the fine details of manuscript painting. Even, I think, Duke Federigo would be impressed; though I grant that both the esthetic and the devotional effect are somewhat swamped by the technological wizardry.

Duke Federigo's passions may be seen in another contemporary setting at the Metropolitan Museum of Art in New York, where his *studiolo,* a finely paneled and decorated small room for the cultivation of the finer pastimes of a duke, has been reconstructed and displayed with immaculate care. Stand in that space and think of his manuscripts and you may begin to sense the grounds for his disdain for print. See then for balance Reeve's study, "Manuscripts Copied from Printed Books," in *Manuscripts from Fifty Years After the Invention of Printing,* ed., J. B. Trapp (London, 1983) 12–20. Derolez on Marcatelli: *The Library of Raphael de Marcatellis, Abbot of St. Bavon's, Ghent, 1437–1508* (Ghent, 1979), and "The copying of printed books for humanistic bibliophiles in the fifteenth century," in *From Script to Book: A Symposium* (Odense, 1986) 140–160.

The quotation from Nicholas of Lyre is taken from A. Minnis and A. B. Scott, *Medieval Literary Theory and Criticism ca. 1100–ca. 1375* (Oxford, 1988) 269. On the practice of destroying manuscripts, see Neil R. Ker, *Pastedowns in Oxford Bindings with a Survey of Oxford Binding ca. 1515–1620* (Oxford, 1954).

Nicholson Baker, in the *New Yorker* (4 April 1994, pp. 68ff., reprinted in his *The Size of Thoughts* [New York, 1996]), attacks the destruction of the old card catalogs in contemporary libraries in just the terms that students of the Renaissance use for the destruction of old manuscripts in the sixteenth century. In both cases, the critic values the old cultural artifact as a thing in itself, full of information that neither producer nor owner ever meant to be there. What was a transparent guide, in the case of the manuscript of Aristotle, to the wisdom of the ancients, and in the case of the card catalog, to the contents of a building's collection, becomes for the humanist scholar an opaque and fascinating cultural phenomenon in its own right. One may accept the validity of the humanist approach while refraining from condemning others for not cramming their attics with every cultural artifact ever produced.

McLuhan has been out of fashion long enough to bear rereading, best in *The Gutenberg Galaxy* (Toronto, 1962) and *Understanding Media* (New York, 1964). Cf. G. Steiner's "On Reading Marshall McLuhan," published 1963 and reprinted in his *Language and Silence* (New York, 1970).

Boethius (ca. 480–525), polymath philosopher and statesman, was author of an impressive stream of books, fell afoul of the Gothic king of Italy Theoderic, and was executed for treason. His *Consolation of Philosophy,* written in prison, is an instructive best-seller: we know nothing of it being read in his own sixth century and little of it after until it became one of the most widely-copied medieval texts in the ninth and later centuries. If we took the canon of great books of western civilization seriously, he would be far better known than he is.

Hyperlink: Who Owns That Idea

On the history of copyright, see Mark Rose, *Authors and Owners: The Invention of Copyright* (Cambridge, Mass., 1993).

The shrine in which the Cathach manuscript was carried may be seen at: *http://ccat.sas.upenn.edu/jod/Picts/cathach.shrine.jpg.* There is a World Wide

Web site for Columba that claims him as the patron saint of "poets, plagiarists, computer pirates, and computer hackers," a dubious distinction.

5. *The Ancients and the Moderns: The Classics and Western Civilizations*

My attention was called to Stalky's implicit message by my colleague Julia Gaisser, whose "The Roman Odes at School: The Rise of the Imperial Horace," *Classical World* 87.5 (1994) 443–456, is a good example of the article in a learned journal that both informs and delights.

On the Battle of the Books in Britain, see J. Levine, *The Battle of the Books: History and Literature in the Augustan Age* (Ithaca, 1991); for the very different but congruent cultural climate in France (with sharp pointers to cultural movements in our own time), see Joan De Jean, *Ancients against Moderns: Culture Wars and the Making of a Fin de Siècle* (Chicago, 1997).

On Bernal's *Black Athena* (New Brunswick, 1988–1991), see the collection of (often polemical) essays, *Black Athena Revisited,* edited by M. R. Lefkowitz and G. M. Rogers (Chapel Hill, 1996).

Simone Weil's essay, "Reflections on the Right Use of School Studies," in her *Waiting for God* (New York, 1951), is very powerful; similar in theme but very different in content is Lawrence Weschler's *Seeing is Forgetting the Name of the Thing One Sees* (Berkeley, 1982), a study of the contemporary minimalist artist Robert Irwin with a title drawn from a line of Paul Valéry.

6. *Augustine Today: Linear Narratives and Multiple Pathways*

Reading Augustine without knowing beforehand what you will find is difficult for those brought up in an ostensibly Christian culture. Some books that put his world in context would include Peter Brown's *Power and Persuasion in Late Antiquity* (Madison, 1992) on the transformation of oratorical culture into bishop-dominated Christianity; also, on that oratorical culture itself, Maud Gleason's *Making Men: Sophists and Self-Presentation in Ancient Rome*

(Princeton, 1995) with illuminating focus on the gender issues usually sublimated when we study such history.

Derrida's memoir is contained in G. Bennington and J. Derrida, *Jacques Derrida* (Chicago, 1993).

7. *The New Liberal Arts: Teaching in the Postmodern World*

The joke about the railroads of the 1950s goes back to Theodore Levitt, "Marketing Myopia," *Harvard Business Review* 38 (July–August 1960) 45–57: "They let others take customers away from them because they assumed themselves to be in the railroad business rather than in the transportation business. The reason they defined their industry incorrectly was because they were railroad-oriented instead of transportation-oriented; they were product-oriented instead of customer-oriented" (p. 45).

The old saw about Mark Hopkins (I learn from an email list after I had quoted it here) goes back to the only American president ever to have served as a professor of classical languages, James A. Garfield, speaking in honor of the longtime nineteenth-century president of Williams College.

9. *Cassiodorus: Or, the Life of the Mind in Cyberspace*

The images from that 1962 issue of *National Geographic* are collected at *http://ccat.sas.upenn.edu/jod/setton.html.* It's consistent with the theme of these essays to divagate a moment on the mysterious superabundance of yellow *National Geographics* in our basements and secondhand book stores. Ours is a society that throws away newspaper and magazines, but has a superstition against throwing away "books." There's a curious reverence for the printed word there, attested by the reluctance to put in the dumpster the "classy" magazines—*National Geographic* but also *American Heritage*—whose printing and binding long resembled those of the printed book. In recent years, more and more magazines, including pornographic magazines, have gone to "perfect binding" to get beyond the limitations of staples, but it's striking to observe that this has not brought with it increased respect for those magazines and preservation in secondhand stores. Can it be that the superabundance of publications in our society has convinced us to observe more rigorously the distinction? *The New Yorker* is still stapled, and survives far less abundantly than *National Geographic.*

Bryn Mawr Classical Review and its sibling, *The Medieval Review,* are online book review journals offering timely assessments of current work in these broad areas. They are specialized in one sense but, because they are free and widely available, have found wide readership among those who simply wish to know more of the ancient and medieval worlds and of the excitement of the best current minds at work in those realms. To browse and search the archives, go to *http://ccat.sas.upenn.edu/jod/bmr.html.*

INDEX